Jan Guzowski

Capillary interactions

Jan Guzowski

Capillary interactions
between colloidal particles at curved fluid interfaces

Südwestdeutscher Verlag für Hochschulschriften

Imprint
Any brand names and product names mentioned in this book are subject to trademark, brand or patent protection and are trademarks or registered trademarks of their respective holders. The use of brand names, product names, common names, trade names, product descriptions etc. even without a particular marking in this work is in no way to be construed to mean that such names may be regarded as unrestricted in respect of trademark and brand protection legislation and could thus be used by anyone.

Publisher:
Südwestdeutscher Verlag für Hochschulschriften
is a trademark of
Dodo Books Indian Ocean Ltd., member of the OmniScriptum S.R.L Publishing group
str. A.Russo 15, of. 61, Chisinau-2068, Republic of Moldova Europe
Printed at: see last page
ISBN: 978-3-8381-2229-8

Zugl. / Approved by: Stuttgart, University and MPI, PhD thesis, 2010

Copyright © Jan Guzowski
Copyright © 2010 Dodo Books Indian Ocean Ltd., member of the OmniScriptum S.R.L Publishing group

Preface

This book is based on my PhD thesis completed at the Max Planck Institute for Metals Research in Stuttgart and at the University of Stuttgart in years 2007 - 2010 under the supervision of Prof. S. Dietrich. Chapters 6 and 7 have been modified as compared to the original thesis and Appendix E has been added. Also, Chapter 7 contains several new figures.

Jan Guzowski
Warsaw, 10.01.2011

Acknowledgements

There are many people who substantially contributed to this thesis in its present shape and here I would like to express my gratitude to them.

First, I would like to acknowledge Prof. Siegfried Dietrich for proposing the subject of the thesis which with time turned out to be more and more interesting and which contributed to my growing enthusiasm towards scientific work in general.

I am also deeply indebted to Mykola Tasinkevych, who was my second supervisor and was vividly supporting me from the very beginning. His competence and patience lead me to understand that if something was difficult to express in words than most probably it had not yet been fully understood (by me). His sense of humor and friendship were a big support for me not only at work.

I gratefully appreciate discussions with Prof. Alvaro Domínguez who provided me with an actual scientific tutorship, particularly on the subject of capillary interactions, and who taught me how important analogies in physics are. I also thank him for refereeing my thesis.

I have to thank all the members of our group at the MPI for a friendly work environment. Particularly, I appreciate many stimulating discussions with Nelson Bernardino, who made me go back to the blackboard or, literally, to the white-board in MPI's kitchen, where numerous new ideas were born.

I also acknowledge fruitful discussions with Martin Oettel, Fabian Doerfler and Ania Maciołek.

Concerning the technical aspects of the software I could always rely on Ken Brakke from Susquehanna University, the author of the source-code to Surface Evolver. His patience in providing me with many helpful tips and tricks has certainly lead to the numerical results of this thesis being obtained in a finite time.

I thank my parents and my brother Krzysztof for their concern and support during over three years of my stay far away from home.

Finally, I would never have done it without love and encouragement of my girlfriend Bogusia.

4

Contents

Preface   1

Acknowledgements   3

1 Introduction   9
   1.1 Particles at fluid-fluid interfaces 10
   1.2 Capillary interactions 13
   1.3 Outline of the thesis 19

2 Capillarity and wetting phenomena   21
   2.1 Equilibrium capillary surfaces 21
      2.1.1 Surface tension and Laplace pressure 22
      2.1.2 Variational approach 24
      2.1.3 Balance of forces acting on the interface 25
      2.1.4 The Young-Laplace equation in the presence of gravity 27
      2.1.5 Linear approximations of the Young-Laplace equation 27
   2.2 Wetting 29
      2.2.1 Wetting energy of a particle at the interface 30
      2.2.2 Wetting and long-range forces 31
      2.2.3 Microscopic approach: density functional theory 32

3 Stability of a particle at a deformable interface   35
   3.1 Free energy functional 36
   3.2 The analytic solution the full Young-Laplace equation 37
   3.3 Mean-force acting on the particle 40
   3.4 Linear theory and renormalization 43
   3.5 Dependence on the system size 45
   3.6 Influence of long-range intermolecular forces 46

4 Capillary interactions   53
   4.1 Free energy of a heavy particle 54
   4.2 Effective interactions 56
      4.2.1 Superposition approximation 58
   4.3 Method of reflections 61

5 Electrostatic analogy 63
5.1 The Poisson equation 63
5.2 Green's function and multipole expansion 65
5.3 Mechanical equilibrium of the interface 67
5.4 Method of images 68

6 Deformations of a spherical droplet 71
6.1 Surface free energy of a spherical droplet. 72
6.2 Expansion in spherical harmonics. 74
6.3 Green's function 75
6.4 Capillary interactions. 77
  6.4.1 Limit of small particles 79
  6.4.2 Explicit formulas for interaction potentials. 81

7 Particles at sessile droplets 83
7.1 Single particle 84
  7.1.1 Free energy functional 84
  7.1.2 Parameterization in terms of spherical coordinates and effective description of the particle 87
7.2 Perturbation theory 89
  7.2.1 Systematic expansion, variation, boundary conditions, and force balance 91
  7.2.2 Green's functions 95
  7.2.3 Free energy 96
  7.2.4 Limit of small particles 96
7.3 Method of images for $\theta_0 = \pi/2$ 99
  7.3.1 Free contact line 100
  7.3.2 Pinned contact line 101
7.4 Many particles 103
  7.4.1 Analytical results for $\theta_0 = \pi/2$ 104

8 Numerical minimization of the free energy 111
8.1 Single spherical particle 112
  8.1.1 Pinned contact line on the substrate 113
  8.1.2 Free contact line on the substrate 115
8.2 Single ellipsoidal particle 116

9 Summary 123

A Calculation of ΔF_{22} 127

B Derivation of the free energy functional 129

C Local approximation of the Green's equation 133

D Calculation of the functions $H(\Omega')$ and $I(\Omega')$ — **137**
D.1 Derivation of the function $I(\Omega')$ 137
D.2 Derivation of the function $H(\Omega')$ 137

E Exact results and linear theory for arbitrary θ_0 — **141**
E.1 Exact results 141
    E.1.1 Free energy functional 141
    E.1.2 Interface profile, surface area and volume 143
    E.1.3 Free energy and mean-force 146
E.2 Comparison with linear theory 149

Bibliography — **150**

Chapter 1

Introduction

For those of us who eat cereals for breakfast the subject of this thesis is closely related to an everyday experience. Small objects, like cereals, floating on the surface of liquid can usually hardly rest. Quite opposite - it is a matter of simple morning experiment to observe their apparently attractive behavior. Some of them will form rafts while the other ones will migrate towards the walls of the dish. Both the formation of clusters and the attraction to the walls can be explained using the concept of surface tension (Vella & Mahadevan 2005). Macroscopically, the surface tension of a liquid-gas interface can be defined as a force acting at a unit length of the three phase contact line, defined as an intersection of the liquid interface with the surface of a floating body. The force is directed tangentially to the liquid interface and such that it tends to diminish the liquid surface area. It opposes external deformations of the interface and therefore the liquid interface effectively behaves like a stretched elastic membrane. (This kind of membrane should not be confused with a biological membrane for which the surface tension can usually be neglected.) As a consequence, heavy objects which normally sink, can float if carefully deposited at the interface. This applies, for example, to metal paper clamps at the water-air interface. Even though the clamps are much denser than water, they are also characterized by large surface to volume ratio, which implies a long three-phase contact line and the resulting force due to surface tension can be large enough to prevent sinking. The attraction between a pair of heavy objects floating at an interface can be understood in analogy to the attraction between a pair of snooker balls placed on a mattress. While one of the balls causes a depression in the surface of the mattress, the second one moves down the depression due to gravity resulting in an apparent attraction.

In the literature, the forces originating from the tension of the liquid interface have been termed capillary forces after the phenomenon of rise of water in a thin capillary (glass tube) immersed in the liquid. The effective interface-mediated interactions between floating objects are often referred to as capillary interactions (Kralchevsky & Nagayama 2001; Domínguez 2010). Scientific interest in capillary interactions arose in the 1940's in view of their possible applications in modeling the behavior of atoms in a crystal. In 1947 Bragg studied rafts of bubbles floating on a soap solution and constituting a two-dimensional lattice (Bragg & Nye 1947). In 1949 Nicolson first studied the interaction between the bubbles quantitatively and indicated that for certain values of system parameters the model replicates pretty well long-range attractions as well as short-range repulsions between atoms (Nicolson 1949). In his analytical approach Nicolson pointed out that the mechanism of attraction between the bubbles is the same as in the case of heavy spheres not wetted by the fluid (which is equivalent to the "mattress" analogy). One of the bubbles causes an inclination of the interface and the other one moves up the incline due to its buoyancy.

In this thesis we investigate capillary interactions, particularly at curved interfaces and under confinement, with applications to particles at sessile droplets, which are of significant practical importance in the industry (for example in inkjet printing). In the remaining part of the introduction we discuss general issues considering particles at interfaces including possible applications in science and technology. Moreover we find it instructive to present an overview of the literature in the field to see how a unified theoretical description of capillary interactions has only recently started to emerge. In fact, it has been a matter of the very last decade that the general mechanism of those interactions, consistent with the fundamental conditions of mechanical equilibrium of the interface, has been understood (Domínguez 2010). In this respect, the theory of interface-mediated interactions seems to be still in its infancy, as compared to the theories of effective interactions in the bulk, like, for example, hydrodynamic interactions (Kim & Karilla 1991).

1.1 Particles at fluid-fluid interfaces

Nowadays, after more than fifty years since the pioneering works of Bragg and Nicolson, particles at interfaces are experiencing a burst of interest owing to development of experimental techniques, such as the video microscopy and the optical tweezers, which enable precise observation and manipulation of individual colloidal particles. Beside numerous technological applications, among which few are mentioned below, the effectively two-dimensional systems of particles are also interesting from the fundamental point of view, because they can be used to obtain direct insight into physical phenomena otherwise hardly accessible experimentally. This includes Casimir forces (Bonn *et al.* 2009; Hertlein *et al.* 2008), quasicrystals (Mikhael *et al.* 2008) or two-dimensional gravitational collapse (Dominguez *et al.* 2010). Other effectively 2D-systems, composed of droplets bouncing from a liquid interface, has been recently demonstrated (Protiere *et al.* 2006; Eddi *et al.* 2009) to exhibit surprising features analogous to particle-wave duality or even quantum tunneling (!).

An important feature of interfacial colloids, which enables their control and manipulation, is that the colloidal particles of sizes varying over many orders of magnitude, from nanometers to millimeters, get irreversibly trapped at fluid-fluid interfaces (Oettel & Dietrich 2008). For example, in the case of a millimeter-sized particle at the water-air interface, the weight and buoyancy of the particle are the forces which compete with the capillary force due to the surface tension. For a micro-particle at the liquid-vapor interface wetting properties take over gravity and equilibrium is determined by the minimum of the surface free energies of the particle-liquid, particle-vapor and liquid-vapor interfaces. If the particle is partially wetted by the liquid, the free energy for a spherical particle is roughly a quadratic function of the particle immersion with a minimum corresponding to the configuration with the particle at the interface (and, more precisely, with the contact angle obeying Young's law). As a consequence, the energy barrier for detachment grows quickly with the particle size and is of the order of $10^3 k_B T$ (at room temperature) for a particle of $10 nm$ in diameter at the air-water interface, and already $10^5 k_B T$ for a particle of $0.1 \mu m$ in diameter. These estimates do not take into account the molecular structure of the interface, in particular thermal fluctuations, however, recent studies based on Monte-Carlo simulations indicate that the trapping mechanism remains effective for particles down to few nanometers in size (Cheung & Bon 2009a; Bresme et al. 2009).

Depending on the typical length scale in an interfacial colloidal system, various physical phenomena become relevant. In the case of particles smaller than a tenth of a millimeter thermal fluctuations can lead to the particle in-plane Brownian motion and to rearrangement of the particle positions. However, when lateral interactions between particles (of not necessarily capillary origin) are strong enough, the particles will rather rest in their local energy minima. In such a situation a two dimensional colloidal crystal can form, such as the one reported by Pieranski in 1980 (Pieranski 1980). With decreasing strength of interactions and increasing random thermal motion one would expect a transition into a fluid phase. In fact, a detailed theoretical analysis of two-dimensional melting as pursued by Kosterlitz & Thouless (1973) and followed by works of Halperin & Nelson (1978) and Young (1978) (KTHNY theory), predicts an additional hexatic phase, intermediate between solid and fluid phases. Surprisingly, the existence of the hexatic phase was for a long time a matter of debate and it was until almost thirty years later that strong experimental evidence supporting KTHNY theory has been given by Zahn & Maret (2000), who observed the three-stage melting using the paramagnetic particles at the air-water interface (see also Zahn et al. 1999). Further modifications of the melting scenario in the presence of anisotropic interactions have also been investigated later in a similar system exposed to an external magnetic field (Eisenmann et al. 2004). Other experiments with two-dimensional colloids have additionally shown that strong lateral confinement of the system can lead to peculiarities like freezing with increasing temperature (Bubeck et al. 1999). An appropriate choice of the external field can even lead to the formation of quasicrystals, as in experiments reported by the group of Bechinger (Mikhael et al. 2008). In these experiments colloidal particles were placed in crossed laser beams, which effectively generated an in-plane external quasi-periodic potential, mimicking the potential experienced by a molecule in a quasicrystal. The transition between a perfect crystal and a quasicrystal, together with an intermediate hexatic phase, was obtained by changing the strength of the external field. All the above examples demonstrate a big advantage of the colloidal systems over the "usual" matter composed of molecules, which is the possibility of precisely tuning the interactions between the particles by controlling the external parameters like the strength of the magnetic field or the laser intensity.

In the case of nanoscopic colloidal particles, fluctuations of the interface themselves can induce remarkable effective interparticle forces (Lehle 2007). The bulk counterpart of those forces is often referred to as a thermodynamic Casimir force (Hertlein et al. 2008) and can be understood in analogy to the Casimir force in quantum electrodynamics, experienced by two parallel metal plates due to quantum fluctuations of the vacuum in a small gap between them (Casimir 1948). In the case of a colloidal system the host fluid plays the role of the fluctuating medium and the range of the interactions grows together with the correlation length as the fluid approaches a critical point. In the case of a fluctuating interface, the height-height correlation function diverges logarithmically with the lateral distance independently of the temperature (Fischer 2004) and therefore the interactions mediated by a fluctuating interface always display a long-range behavior. However, the strength of those interactions is of the order of k_BT (Lehle et al. 2006) and therefore, at room temperature, they could play a significant role only for sufficiently small particles (nanoparticles). In such a case, again, the effective interactions could be tuned by changing a single parameter, this time the actual temperature.

The equilibrium shape of a free interface obeys the condition of a constant curvature, which is expressed by the Young-Laplace equation

$$2\gamma H(\boldsymbol{r}) = \Delta p, \tag{1.1}$$

where γ is the surface tension, $H(\boldsymbol{r})$ is the mean curvature of the interface at point $\boldsymbol{r}$ and Δp is the pressure change across the interface. For small deformations u of a flat interface, the curvature $2H$ can be approximated in the Monge parameterization by minus the two-dimensional Laplacian of the deformation. Then the shape of the interface is governed by the two-dimensional Poisson equation,

$$-\gamma \nabla_\parallel^2 u(\boldsymbol{x}) = \Delta p, \tag{1.2}$$

where $\nabla_\parallel = \boldsymbol{e}_x \partial_x + \boldsymbol{e}_y \partial_y$ and $\boldsymbol{x}$ is the in-plane coordinate. Eq. (1.2) is formally equivalent to the two-dimensional electrostatics or two-dimensional gravity (assuming Newtonian gravity) with the deformation u playing the role of a potential and Δp the role of a charge (or mass) distribution. In this analogy a pointlike particle deforming the interface can be treated as a unit "capillary charge" or a unit "capillary mass" and one could imagine that, as such, colloidal particles could possibly even be used to experimentally study the gravitational collapse in 2D-systems (Dominguez et al. 2010). Of course the tempting picture of "cosmology in a Petri dish" has its obvious limitations, like the fact that the dynamics of colloidal particles is governed rather by the Langevin equation and not by Newton's equations of motion.

The above examples show that particles at interfaces provide an easily accessible model system for studying laws of physics. On the side of direct applications, two-dimensional confinement of the particles has practical importance for precise manipulation of the particles, which could be used, e.g., for patterning of solid substrates with desired two-dimensional structures (Denkov et al. 1993). Some experimental techniques combine capillary forces with the use of sculpted substrates as templates in order to assemble particles into regular arrays (van Blaaderen et al. 1997; Aizenberg et al. 2000; Cui et al. 2004).

Thermodynamically stable structures can also form spontaneously at the interfaces owing exclusively to the interactions between the particles (i.e., without a template). Those interactions may not only be considerably modified in the neighborhood of the interface, as compared to their counterparts in the bulk, but, as already mentioned, they can even emerge exclusively due to the presence of the interface. This is the case for macroscopic bubbles, as those studied by Nicolson, which do not exhibit long-range interaction in the bulk, as well as for colloidal particles forming a crystal at water-oil interface in Pieranski's experiment. However, in the latter case, the micron-sized particles always remain separated by a few particle diameters, which indicates the existence of a repulsive component of the interaction potential. In fact, small spherical particles, for which gravity and other forces are negligible, can be in equilibrium with a perfectly flat, i.e., undeformed interface, which rules out capillary interactions. Instead, as noticed by Pieranski and later quantitatively studied by Hurd (Hurd 1985), as soon as the adjacent fluids have different dielectric constants, the interface breaks the symmetry in the ion distribution around each particle which leads to an effective electrostatic dipolar repulsion. Macroscopically, as a result of this repulsion particles arrange themselves in a regular hexagonal lattice.

In the last decade many authors reported on spontaneous formation of colloidal two-dimensional mesostructures, including clusters, stripes and voids (Ruiz-Garcia et al. 1998; Ghezzi & Earnshaw 1997; Ghezzi et al. 2001; Sear et al. 1999). These observations suggest more complicated form of the effective interaction potentials than just a repulsion, in particular a presence of both repulsion and attraction. Indeed, as shown in recent numerical studies (Archer 2008), the interaction potential in the form of a hard-core repulsion together with an attractive part at small distances and again a repulsive part at large distances can yield a rich variety of equilibrium structures, depending on the surface coverage (density). At lower densities particles form arrays of circular clusters, but at higher densities series of elongated parallel stripes are observed. Further increasing the density leads to formation of arrays of voids within liquid-like regions. Other numerical simulations of a two-dimensional system of a finite-number of particles with strong lateral confinement have shown that proper tuning of the interaction parameters can yield even more kinds of equilibrium structures (Liu et al. 2008). However, the emergence of the potentials used in simulations from the underlying microscopic degrees of freedom (like ion distribution or deformations of the interface around the particles) is in many cases still obscure. Moreover, the effects of possible interfacial contaminations may easily be mistaken for pair-interactions (Fernandez-Toledano et al. 2004).

1.2 Capillary interactions

In general, the effective interactions between colloidal particles at interfaces emerge as a consequence of the dependence of the free energy on the separation between the particles (Oettel & Dietrich 2008). At the length scale of millimeters the gravitational energy dominates (Kralchevsky & Nagayama 2000), whereas for micrometer-sized systems the interactions are driven by the changes in the surface free energy (Oettel et al. 2005b). At the nanoscale, the presence of particles imposes constraints on the thermal fluctuations of the interface and as a consequence the statistical sum over all possible configurations of the interface (and so the free energy) depends on the separation of the particles (Lehle et al. 2006). In this thesis we focus on the capillary interactions between particles of micrometer-sizes, which are small enough such that their buoyancy can be safely neglected, but which on the other hand are big enough such that the thermal fluctuations can be neglected. Thus, the deformation of the interface is induced only by the wetting properties or other constraints on the shape of the interface at the particle surfaces (like pinning of the contact line due to roughness). As mentioned in the previous Section, small deformations of an initially flat interface obey the same governing equation as a two-dimensional electrostatic potential and thus decay logarithmically with the distance (this holds only at the length scales much smaller than the capillary length, which is usually the case because the latter is of the order of millimeters, such that it exceeds the particle size by orders of magnitude). Within this electrostatic analogy a pointlike perturbation caused by an external force f perpendicular to the interface can be treated as a "capillary charge" $Q = f$ (Domínguez et al. 2008a; Domínguez 2010). The interaction potential between two arbitrary charges Q_1 and Q_2 separated by lateral distance d is then analogical to Coulomb's law in electrostatics,

$$V_{cap}(d) = \frac{Q_1 Q_2}{2\pi\gamma} \ln\left(\frac{d}{L}\right), \tag{1.3}$$

where L is the large distance cut-off (which might be for example the system size or the capillary length). As one can see from Eq. (1.3) the analogy with electrostatics is exact up to reversal in the sign of the interactions. Therefore, two like capillary charges attract each other, while the opposite is true in electrostatics. Colloidal particles may carry a capillary charge or a capillary dipole or even higher multipoles depending on how they interact with the interface. A capillary monopole corresponds to a particle pulled or pushed by an external force (this might be, for example, the weight or the buoyancy of the particle), whereas a capillary dipole corresponds to an external torque acting on the particle. In absence of external agents the deformation around a particle corresponds to a capillary quadrupole. Furthermore, mechanical equilibrium of the interface itself implies that all the external forces and torques acting at the interface must cancel each other. Thus, an external force or torque exerted at the interface by the particle must be balanced by the force and torque exerted by the walls of the container. This complies with the long-range character of capillary interactions, expressed in Eq. (1.3), and means that the boundary conditions at the walls in general cannot be neglected.

The logarithmic behavior of monopole-monopole capillary interactions also implies that, as soon as there is a mechanism by which the monopoles are induced, they should, at large distances, overcome electrostatic dipolar repulsion as those in Pieranski's experiment. The formation of freely floating crystallites composed of a few particles (up to 7) at a water droplet suspended in oil, reported in a recent experiment (Nikolaides et al. 2002), indeed indicates the presence of a long-range attraction between particles at the interface. The authors proposed a simple explanation in which the particles are pulled towards the water phase by the electrostatic forces due to asymmetric ion distribution around the particle due to the different dielectric constants of the adjacent fluids. Under this assumption, the particles carry not only electrostatic dipoles,

1.2. CAPILLARY INTERACTIONS

as in Pieranski's reasoning, but also, due to a "pulling" force in the normal direction, capillary monopoles. However, the authors assumed that the interface is flat and unbounded, which contradicts the condition of mechanical equilibrium.

Indeed, as indicated by Megens & Aizenberg (2003), the presence of an unbalanced force (capillary monopole) in a mechanically isolated system (no walls) would be inconsistent with Newton's third law. Instead, the latter authors suggested that in the presence of the electrostatic pressure, decaying as r^{-6} with the lateral distance r from the particle, the interaction potential should be attractive and decay also like r^{-6}. A similar reasoning was subsequently presented by Foret & Würger (2004), who obtained the same power law, but with an opposite sign. Their calculations were based on a regularization of the interface deformation at $r = 0$ and a superposition approximation for the electrostatic pressure in the presence of two particles. However, it has been later noted by Domínguez et al. (2005) and Oettel et al. (2005a) that the latter assumption is not correct because the electrostatic pressure, proportional to the electrostatic field squared, is not additive. Instead, one should use the additivity of the electrostatic field, which finally gives an attractive r^{-3} behavior. This leads to the conclusion, that the total interparticle potential including the direct dipolar electrostatic repulsion, governed also by the power law r^{-3}, cannot exhibit a minimum, unless the Debye screening length in water is comparable with the size of the particles. In such a case a shallow minimum with a depth of several $k_B T$ could be expected (Domínguez et al. 2007b). However, its occurrence would depend sensitively on the precise values of several system parameters. Particularly, it would correspond to relatively large electric field and as a consequence to a large deformation of the interface around the particles for which the perturbation theory would actually loose its validity. Apart from that, it must be noted that in the mentioned experiment, the droplets of water, at which the particles were observed, were in fact attached to a solid plate. Therefore, any quantitative or maybe even qualitative predictions cannot be satisfactory as long as the curvature of the interface and the boundary conditions at the plate have not been taken into account. This problem has not yet been comprehensively studied in the literature and it is one of the issues addressed in this thesis.

When external forces on the particles can be neglected, the capillary interactions are ruled by the lowest non-vanishing moment, which is quadrupole, and therefore they become anisotropic. The effects connected with the anisotropic nature of capillary interactions has been demonstrated in numerous experiments and simulations using non-spherical particles (Bowden et al. 1997; Brown et al. 2000; Stamou et al. 2000; Loudet et al. 2005; van Nierop et al. 2005; Madivala et al. 2009). Experimental techniques enable fabrication of microscopic particles of many different shapes including the shape of a bent disk, able to pin the interface at its perimeter (Brown et al. 2000). In the case of ellipsoidal particles, characterized by the contact angle different than $\pi/2$, the three-phase contact line must undulate in order to meet the surface of the particle at a constant angle given by Young's law. The quadrupolar deformation field around micrometer ellipsoids has been directly measured for particles of various aspect ratio (Loudet et al. 2006). In the case of many particles the anisotropy of the interactions leads to complicated energy landscapes with large number of local minima resulting in a variety of possible metastable structures (Fournier & Galatola 2002; van

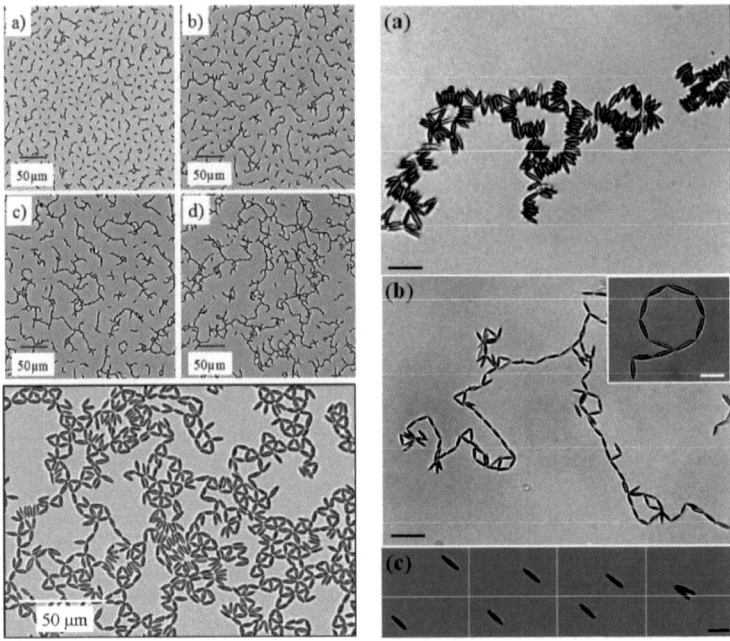

Figure 1.1: Left panel: equilibrium chains of self-assembled ellipsoids at the oil-water interface arranged tip-to-tip and side-to-side, depending on the surface chemistry of the particles (reprinted with permission from Loudet et al. 2005, Copyright 2010 by American Physical Society). Chains of particles bend due to many-body capillary interactions; right panel: time evolution of a system of ellipsoidal particles at the decane-water interface with the final local triangular and flower-like structures (reprinted with permission from Madivala et al. 2009, Copyright 2010 by American Chemical Society).

Nierop et al. 2005). Moreover, multipole moments higher than quadrupoles become important at very small separations and can lead, for example, to the formation of linear chains of ellipsoids arranged tip-to-tip (Stamou et al. 2000; Fournier & Galatola 2002). Numerical studies (Fournier & Galatola 2002) incorporating three and more particles have further shown an intriguing many-body effect, that the particles in such a chain meet at a non-trivial angle forcing the chain to bend (see Fig. 1.1), which agrees with the experimental observations (Loudet et al. 2005). At higher concentrations, the equilibrium structures seem to depend strongly on the aspect ratio of the ellipsoids. For nearly spherical particles, the hexagonal positional order is preferred, characteristic for a system with isotropic interactions, together with the simultaneous orientational order of herringbone type (van Nierop et al. 2005). For larger aspect ratios the ellipsoids arrange tip-to-tip and form locally triangular and flower-like structures (Fig. 1.1),

1.2. CAPILLARY INTERACTIONS

which under further compression rearrange into local side-to-side configurations. Single particles may also flip such that their longer axes point in the direction normal to the interface, which allows for a more closely-packed structure (Madivala et al. 2009).

In the case of particles at the surface of liquid droplets, non-trivial global curvature of the interface can also influence effective interactions. However, the issue of particles at curved interfaces goes beyond the mentioned experiments with particles at oil-in-water droplets. For example, lipid bilayers forming boundaries of living cells are often decorated with inclusions, which may induce variations in thickness of the bilayers and thus exhibit effective interactions (Israelachvili 1977). Another example comes from the world of insects: an inclination of the meniscus can pose an obstruction for small animals to get from water to land or vice-versa (Hu & Bush 2005). To overcome this difficulty, certain water-climbing insects developed an ability to adjust their body posture depending on meniscus curvature such that they get dragged by capillary forces towards the contact line, against gravity. Other insects use the lateral capillary forces to assemble themselves into large colonies of 50-100 animals (see Fig. 1.2) and to travel between colonies. On the side of technological applications, colloidal particles are known to stabilize emulsions by self-assembling at the droplet surfaces and forming so called Pickering emulsions (Pickering 1907). Closed packed spherical shells of such particles might have potential use in drug delivery in medicine (Dinsmore et al. 2002). The particles at droplet surfaces can also be used to study basic problems like correlations (Chavez-Paez et al. 2003; Viveros-Mendez et al. 2008) on a sphere or topological defects in spherical two-dimensional crystals (Bausch et al. 2003).

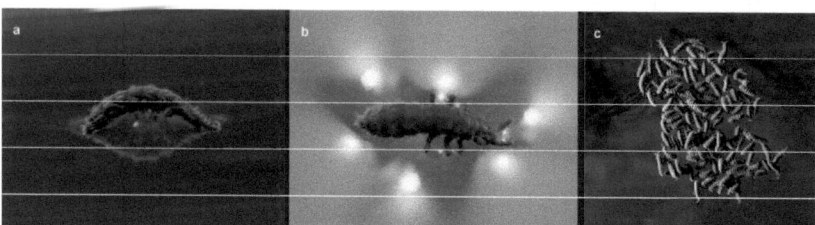

Figure 1.2: Meniscus-climbing insects (here, *Anurida maritima*) can adjust their body posture in order to be dragged by surface tension up the water meniscus. Due to the capillary forces they can also assemble themselves into colonies or even move between the colonies. Image courtesy of John Bush, MIT.

In this thesis we study the influence of interface curvature and non-spherical particle shapes on capillary interactions at the surface of a sessile droplet, which is a simple model system for studying static self-assembly (Whitesides & Grzybowski 2002) in confined geometries. We provide a theoretical basis for addressing the question of how the properties of the individual particles, together with the geometry of the system, influence the final equilibrium structure, which can be identified with a bottom-up approach to self-assembly (Pelesko 2007). Beside the scientific interest, colloidal droplets could be of practical importance for assembling structures on solids. If spilled on a table, small droplets of coffee leave circular deposits of particles after evaporation. This so-called coffee-ring effect is driven by the evaporation of liquid from the droplet, which results in a flux of particles towards the contact line (Deegan et al. 1997). This dynamical non-equilibrium phenomena has been recently thoroughly studied (Deegan et al. 1997; Dietzel & Poulikakos 2005; Bigioni et al. 2006; Kuncicky & Velev 2008; Murisic & Kondic 2008). However, not much is known about particles at the surface (or inside) non-evaporating droplets, which may also exhibit pattern formation by finding their equilibrium positions corresponding to the free energy minima. Some experimental and theoretical efforts has been devoted to explain the interactions between the particles close to the three phase contact line at the substrate, but the effects of the curvature and the finite volume of the droplet on the interactions were usually neglected (with one exception, see Sangani et al. 2009), due to small interparticle separations as compared to the droplet size (Helseth & Fischer 2003; Helseth et al. 2005). Interactions between particles protruding from a spherical liquid film were studied by Kralchevsky et al. (1995). In special cases (cork-shaped particles, pinned contact lines) the authors obtained non-monotonic interaction free energies leading to an attraction at small separations and repulsion at large separations (which might possibly lead to clusterization (Archer 2008)), but the physical mechanism responsible for this effect has not been explained.

If the particles are trapped at the interface of a microscopic sessile droplet, the surface energies play a crucial role. Then, besides mutual capillary interactions each particle is subjected to an effective free energy landscape induced by the presence of the substrate. This kind of capillary effect may influence the final equilibrium structures, and one cannot rule out that it could have been the case in the experiments reported by Nikolaides et al. (2002). The possible importance of the effects of curvature on capillary interactions has been mentioned by Oettel et al. (2005b), who derived the equation for the axisymmetric shape of a slightly deformed spherical sessile droplet with a particle at its apex, taking into account the volume constraint and a certain type of boundary conditions at the substrate. In a subsequent publication (Domínguez et al. 2005) the same authors indicated that the logarithmic term in the deformation field around the particle can only occur in the situation when the droplet is not mechanically isolated (for example, if it is supported by a substrate). The effects of curvature on capillary interactions have been also studied by Würger (2006a;b), who claimed that a logarithmically decaying deformation field can occur due to the curvature itself even at a mechanically isolated droplet. However, as indicated by Domínguez et al. (2007a) Würger's reasoning did not correctly resolve the force balance on the droplet, and the logarithm appeared only as an artifact of an unbalanced force.

Our approach attributes mostly to the work of Morse & Witten (1993), who considered weakly compressed emulsions rather than interfacial colloids. In such emulsions droplets get deformed due to the contact with other droplets and then the deformation energy can be calculated by means of Green's function by approximating the areas of contact by pointlike forces. In this thesis we argue that the point-force approximation is also valid for spherical colloidal particles and the mechanism of their attachment to the interface turns out to be of secondary importance. In such an approach the free energy landscapes as well as pair-interactions can be calculated by using Green's

1.3 Outline of the thesis

The thesis is composed as follows. In Chapter 2 we introduce the Young-Laplace equation and discuss its general solutions in various coordinate systems, including spherical coordinates. In the same Chapter we discuss shortly the wetting phenomena and their relevance for particles at interfaces. In Chapter 3 we study the surface free energy of a particle at a deformable, initially flat interface enclosed in a cylindrical container (neglecting the influence of gravity). The exact solutions of the full nonlinear Young-Laplace equation obtained for the case of an axisymmetric interface serve as a check for a renormalized linear theory for small deformations of the interface. We show that both theories predict, with remarkable mutual agreement, metastable branches of the free energy as a function of the height of the particle above the interface. We find that the external vertical force, needed to fix the particle at a given position, uniquely determines the shape of the interface. Next, we study the case of two heavy particles at an unbounded interface in presence of gravity and explain the mechanism of capillary interactions at a flat reference interface in Chapter 4. In Chapter 5 we investigate the analogy between capillarity and electrostatics, in which the external vertical force acting on the interface can be interpreted as a "capillary charge" and thus a particle of an arbitrary shape can be replaced by a proper capillary charge distribution. Small deformations of a spherical droplet subjected to an arbitrary pressure field are studied in Chapter 6. The analytical results obtained there enable approximate calculations of the interaction free energy between particles of various shapes at a spherical interface, in the limit of small particles. In Chapter 7 the system composed of a single spherical particle at a *sessile* droplet is studied in detail. The boundary conditions at the substrate are expressed in terms of small deformations of a reference cap-spherical droplet shape. The free energy landscape for the particle pulled radially by an external force is calculated analytically in the point-force approximation in the special case $\theta_0 = \pi/2$ for a free and for a pinned contact line at the substrate. In both cases pair-potentials are also carried out analytically. Results of numerical minimization of the free energy for a spherical as well as for an ellipsoidal particle and comparison with the approximate analytical results are presented in Chapter 8. Chapter 9 provides a summary of the thesis.

Chapter 2

Capillarity and wetting phenomena

This Chapter introduces the basic concepts of capillarity and wetting phenomena which underly capillary interactions between particles at fluid interfaces. In Sec. 2.1 we study the equilibrium shapes of fluid interfaces governed by surface tension and in Sec. 2.2 we explain how a particle may get trapped at the interface due to its wetting properties. Finally we introduce the simplest mean-field version of the density functional theory (Evans 1979), in terms of which the problem of the particle at the interface can also be addressed at a microscopic level in terms of intermolecular forces (for a very recent assessment in this topic see Hopkins et al. 2009).

2.1 Equilibrium capillary surfaces

Basic ideas concerning the shape of fluid interfaces date back to the the beginning of XIX'th century, to the works of Young (Young 1805) and Laplace (de Laplace 1805 and 1806), who considered the problem of the rise of water in a very thin vertical glass tube dipped into the liquid. This phenomenon was then in an apparent contradiction with the laws of hydrostatics and could not be explained on the ground of Newtonian gravity. Indeed, one can easily convince oneself that the interaction of liquid with glass in this case does not depend on the thickness of the glass. The observed height of the liquid in the tube does neither depend on kind of the liquid. The solution of this problem was one of the first examples of a link between a macroscopic phenomena and the underlying discrete structure of matter. (In general this kind of reasoning is nowadays prescribed to statistical physics, however, at the beginning of XIX'th century, the molecules were assumed not to exhibit significant motion in equilibrium, and it was not until the works of Boltzmann and Gibbs that the statistical description of matter actually emerged.) The basic concept of the early theory of capillarity was the assumption that the molecules of liquid attract each other. As a consequence, removing a group of particles from the bulk liquid, costs energy (Rowlinson & Widom 2002). This energy can be associated with creating two liquid surfaces: the surface of the droplet and the surface of the cavity in the bulk liquid. The surface energy, expressed at the unit area of the created surface, has been called the surface tension. In this Section we analyze the macroscopic consequences of the surface tension such as the equilibrium shape of the liquid interfaces and the capillary force.

2.1.1 Surface tension and Laplace pressure

In this Subsection we present an explanation of the rise of liquid in a capillary tube (for a classical presentation of this problem see Rowlinson & Widom 2002) which includes a simple intuitive derivation of the Young-Laplace equation (Eq. (2.5)).

Macroscopically, surface tension has an interpretation of a force acting at a unit length of a boundary of the liquid surface. In order to illustrate this, imagine a liquid film in a rectangular wire frame with one side, of length L, being free to move and held by force f, which is needed to balance the tension in the two-sided film. Displacing this boundary by the distance δx requires the work $f \delta x = \gamma \delta S$, equal to the change in the surface energy, where $\delta S = 2L \delta x$ is the change in the area of both surfaces of the film. Thus the capillary force equals $f = 2\gamma L$.

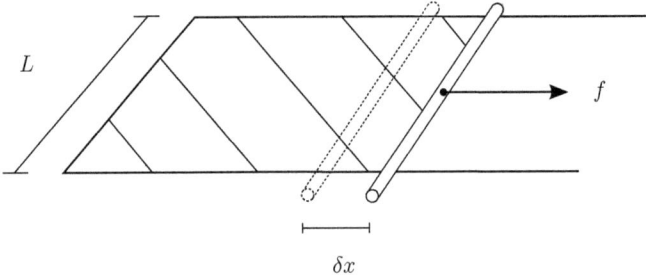

Figure 2.1: Liquid film held in a rectangular wire frame exerts a capillary force on the rod, counterbalanced by the external force f. The work of the external force $f \delta x$ done in displacing the rod is equal to the increase of the surface free energy of the film by $2\gamma L \delta x$. Thus the capillary force equals $f = 2\gamma L$.

To explain the rise of liquid in a capillary tube not only the surface tension but also the curvature of the surface plays a crucial role. Taking a tube of a diameter exceeding (in the case of water-air interface) $2 - 3$ mm the effect will disappear, because the interface becomes flat. In fact, internal pressure of liquid depends on the curvature of the interface and it is the resulting difference in pressure that pulls the liquid up the tube.

In the mechanical interpretation discussed above, one can treat the liquid surface as an elastic membrane subjected to the tension γ. In order to see how this tension leads to the shift in pressure, consider a spherical gas bubble of radius R inside a body of liquid. For the bubble to exist, the tension of the surface, which tends to squeeze the bubble must be balanced by an excess pressure Δp. In equilibrium the virtual work associated with a virtual displacement δR of the radius must vanish, which can be written as

$$\gamma \delta S - \Delta p \delta V = 0. \qquad (2.1)$$

2.1. EQUILIBRIUM CAPILLARY SURFACES

where $\delta S = 8\pi R \delta R$ is the change in surface area and $\delta V = 4\pi R^2 \delta R$ is the change in volume, such that the excess pressure, called the Laplace pressure, is given by

$$\Delta p = \frac{2\gamma}{R}. \tag{2.2}$$

For a very thin tube one can assume that the interface is a portion of a sphere. In such a case the curvature must increase (and so the height of liquid in the tube) with decreasing diameter of the tube as long as the contact angle θ between the interface and the glass surface is kept constant. The idea of constant contact angle has been first proposed by Young (Young 1805) and it can also be justified using the concept of surface tensions. Indeed, mechanical equilibrium demands that the surface tension γ of the liquid-gas interface acting on the three-phase contact line is balanced by the surface tensions γ_{sl} and γ_{sg} of the glass-liquid ("solid-liquid") and glass-vapor ("solid-gas") interfaces (see Fig. 2.2). Approximating the region of contact by a liquid wedge we find the relation

$$\cos\theta = \frac{\gamma_{sg} - \gamma_{sl}}{\gamma}, \tag{2.3}$$

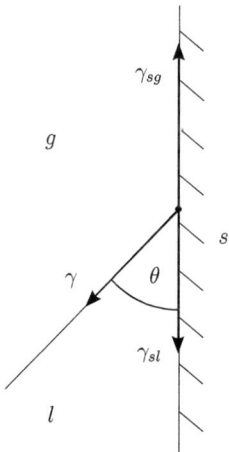

Figure 2.2: Section of a liquid-gas (l, g) interface in contact with a wall (in general a solid substrate s). The arrows schematically correspond to forces exerted by the surface tensions γ_{sl}, γ_{sg} and γ at a unit length of the contact line (perpendicular to the drawing). Young's law in Eq. (2.3) can be understood as a condition of balance of those three forces.

which states that the contact angle is a function only of the surface tensions and not, particularly, of the shape of the tube. Given the contact angle $\theta < \pi/2$ of water and glass, the meniscus inside a capillary of diameter $2R$ must be convex, and its radius of curvature equals $R/\cos\theta$. The resulting Laplace pressure $2\gamma\cos\theta/R$ is balanced by the hydrostatic pressure provided the liquid inside a capillary rises to a height h, such that

$$h = \frac{2\gamma\cos\theta}{\rho g R}. \tag{2.4}$$

The law of Laplace in Eq. (2.2) holds in fact not only for the spherical interface, for which the principal radii of curvature are equal, but also for interfaces of arbitrary curvature given by the two different principal radii of curvature R_1 and R_2. It can be written in a general form as

$$\Delta p = \gamma(R_1^{-1} + R_2^{-1}), \tag{2.5}$$

which is known as the Young-Laplace equation and relates the pressure change across the interface to the curvature of the interface. The concept of a constant contact angle and the relation between the pressure and the curvature of the interface, together with the laws of hydrostatics, form a basis sufficient to calculate the shape of a liquid surface bounded by walls of arbitrary geometry. For many practical purposes it is often more convenient to use an equivalent, variational approach, which was proposed by Gauß (Gauss 1830), following the works of Young and Laplace.

2.1.2 Variational approach

Consider a surface above the xy-plane such that it can be parameterized in Cartesian coordinates as

$$z = z(x,y). \tag{2.6}$$

The infinitesimal element of the surface area d^2S corresponding to $dxdy$ is given by $dxdy|\partial_x \boldsymbol{r} \times \partial_y \boldsymbol{r}|$ where

$$\boldsymbol{r}(x,y) = \begin{pmatrix} x \\ y \\ z(x,y) \end{pmatrix}, \tag{2.7}$$

so that d^2S can be written as $d^2S = dxdy\, s(x,y)$ with s given by

$$s(x,y) := \sqrt{1 + \left(\nabla_\parallel z\right)^2}. \tag{2.8}$$

The corresponding infinitesimal element of the volume enclosed between the interface and the xy-plane reads $dxdy\,z$. The equilibrium shape of the liquid surface minimizes a functional $\mathcal{F}$ being the sum of surface energy γS for a prescribed volume V of liquid,

$$\mathcal{F} = \gamma S - \lambda V = \int dx \int dy\,(\gamma s - \lambda z) \tag{2.9}$$

where $-\lambda$ is the Lagrange multiplier. On the other hand, treating V as an extensive variable (and not a functional of $z(x,y)$), one has $\lambda = -\partial \mathcal{F}/\partial V$, so that λ can be identified with the internal pressure of the liquid.

2.1. EQUILIBRIUM CAPILLARY SURFACES

Minimizing $\mathcal{F}$ with respect to z yields the Euler-Lagrange equation in the form

$$-\gamma \operatorname{div} \frac{\nabla_\| z}{\sqrt{1+(\nabla_\| z)^2}} = \lambda, \qquad (2.10)$$

It can be shown by explicit calculation by means of the differential geometry (Finn 1986), that the left hand side equals the sum of reciprocal radii of curvature of any two orthogonal curves at the point of their intersection. Thus, identifying λ with the pressure Δp we recover the Young-Laplace equation (2.5). It can also be shown that, incorporating the wall surface energy into $\mathcal{F}$ would yield Young's law (Eq. (2.3)). The proof in the case of an arbitrary geometry of the substrate can be found in the work of Finn (1986). In general, Young's law does not hold in the presence of substrate heterogeneities which prevent the relaxation of the contact angle to its equilibrium value. This case will be discussed on the example of a sessile droplet with the contact line pinned at the substrate in Chapter 7.

2.1.3 Balance of forces acting on the interface

The divergence structure of the left hand side of Eq. (2.10) can be used to derive the balance of forces acting on arbitrary piece of interface S. Integrating over the projection A of S onto the xy-plane and applying a two-dimensional version of the divergence theorem yields

$$-\gamma \oint_{\partial A} dl_\| \frac{\boldsymbol{n} \cdot \nabla_\| z}{\sqrt{1+(\nabla_\| z)^2}} = \lambda A, \qquad (2.11)$$

where $\boldsymbol{n}$ is a unit vector in the XY-plane normal to the infinitesimal element $dl_\|$ of the boundary ∂A. The lhs of Eq. (2.11) represents minus the total capillary force acting on the boundary ∂S in the vertical direction, which can be seen by the following reasoning. We note that the capillary force acting at an infinitesimal element dl of the contour ∂S equals $\boldsymbol{e}_t \gamma \, dl$, where $\boldsymbol{e}_t$ is a unit vector tangent to the interface, perpendicular to ∂S and pointing outside S (see Fig. 2.3), and reads

$$dl \boldsymbol{e}_t = \frac{\partial_x \boldsymbol{r} \times \partial_y \boldsymbol{r}}{|\partial_x \boldsymbol{r} \times \partial_y \boldsymbol{r}|} d\boldsymbol{r} \times \boldsymbol{e}_n, \qquad (2.12)$$

where $\boldsymbol{r}$ is the radial vector running on ∂S and $d\boldsymbol{r}$ is directed along ∂S; $\boldsymbol{e}_n$ is a unit vector normal to S at $\boldsymbol{r}$,

$$\boldsymbol{e}_n = \frac{1}{\sqrt{1+(\nabla_\| z)^2}} \begin{pmatrix} -\partial_x z \\ -\partial_y z \\ 1 \end{pmatrix}. \qquad (2.13)$$

Hence, minus the total capillary force in the z direction equals

$$-\gamma \oint_{\partial S} dl \, \boldsymbol{e}_t \cdot \boldsymbol{e}_z = \gamma \int \frac{-dy \partial_x z + dx \partial_y z}{\sqrt{1+(\nabla_\| z)^2}}. \qquad (2.14)$$

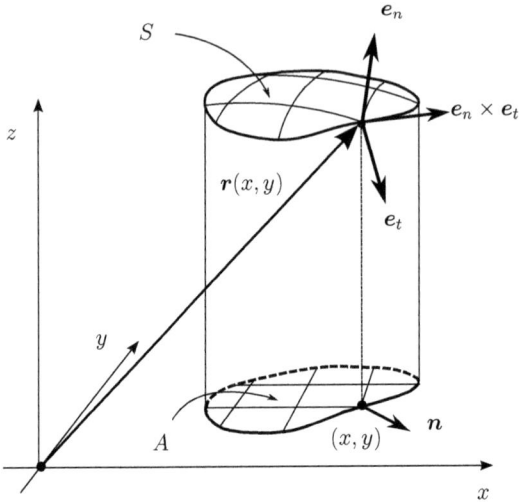

Figure 2.3: Piece of the interface S in a Monge parameterization $z = z(x,y)$ and its projection A onto the xy-plane.

By noting that
$$-dy\boldsymbol{e}_x + dx\boldsymbol{e}_y = -dl_\parallel \boldsymbol{n}, \tag{2.15}$$
we can rewrite the rhs of Eq. (2.14) such that it yields the lhs of Eq. (2.11). Moreover, from the definition of the projected area A we have
$$A = \int_S d^2S\, \boldsymbol{e}_n \cdot \boldsymbol{e}_z. \tag{2.16}$$
Thus, Eq. (2.11) is equivalent to
$$-\left(\gamma \oint_{\partial S} dl\, \boldsymbol{e}_t + \int_S d^2S\, \lambda \boldsymbol{e}_n\right) \cdot \boldsymbol{e}_z = 0. \tag{2.17}$$
However, the direction z has been chosen arbitrarily, and therefore the expression in brackets, which is explicitly invariant with respect to the parameterization of S, must vanish. We obtain
$$-\gamma \oint_{\partial S} dl\, \boldsymbol{e}_t = \int_S d^2S\, \lambda \boldsymbol{e}_n, \tag{2.18}$$
which is an integral representation of the Young-Laplace equation (without gravity) and expresses the condition of mechanical equilibrium of an arbitrary piece S of the interface in which the force due to pressure λ integrated over S (lhs) equals minus the total capillary force acting on ∂S (rhs).

2.1.4 The Young-Laplace equation in the presence of gravity

In the presence of gravity the contributions to the surface energy come not only from the surface tension but also from the weight of the liquid enclosed under the interface. The total pressure change Δp across the interface can be written as a sum of the constant term λ and the hydrostatic pressure $-\Delta \rho g z$, where $\Delta \rho$ is the density difference between the lower and the upper fluid phases. The corresponding Young-Laplace equation can be written in the form (Langbein 2002)

$$-\gamma \operatorname{div} \frac{\nabla_\| z}{\sqrt{1+(\nabla_\| z)^2}} = \lambda - \Delta \rho g z. \tag{2.19}$$

The gravitational field directed along the z-axis breaks the isotropy and the force balance has a different form depending on the direction. Particularly, choosing the z-direction, one obtains an equation analogical to Eq. (2.18):

$$-\gamma \oint_{\partial S} dl\, \boldsymbol{e}_t \cdot \boldsymbol{e}_z = \int_S d^2 S\, (\lambda - \Delta \rho g z) \boldsymbol{e}_n \cdot \boldsymbol{e}_z, \tag{2.20}$$

Obviously, the rhs of Eqs. (2.19) and (2.20) do depend on the position of the reference plane. If the volume of liquid, defined as the volume enclosed between the interface and the reference plane, is not prescribed then one can actually always choose the reference plane such that $\lambda = 0$ (see Eq. (2.19)). However, if the liquid volume is given *a priori*, then λ must be determined from the volume constraint.

2.1.5 Linear approximations of the Young-Laplace equation

We note that flat and spherical interfaces both spontaneously occur during the processes of phase separation. For example, during condensation of liquid from the gas phase, initially small amounts of liquid form perfectly spherical droplets, which finally merge to form a single flat ($g \neq 0$) horizontal interface, separating heavier and lighter fluid below and above the interface, respectively. On the other hand, in the absence of gravity, the equilibrium shape of the interface is uniquely determined by the shape of the container (Finn 1986). Here, we study the Young-Laplace equation (2.19) in the limit of small deformations of a reference flat interface.

Small deformations of a flat interface

Consider a small deformation $u(x, y)$ of an initially flat interface. In the limit $|\nabla_\| u| \ll 1$ the Young-Laplace equation (2.19) with $\nabla_\| z = \nabla_\| u$ can be approximated up to the linear terms in $\nabla_\| u$ by

$$-\nabla_\|^2 u + q^2 u = 0, \tag{2.21}$$

where $q = \lambda_c^{-1} = \sqrt{\Delta \rho g / \gamma}$ is the inverse capillary length (λ_c). Without loosing generality we have put the Lagrange multiplier λ equal to zero due to the fact that the liquid volume is not fixed. A non-vanishing λ would contribute only a constant to the deformation u, which could then be interpreted as a translation of the reference plane without any change in the shape of the interface. The equation (2.21) can be separated in polar coordinates (r, ϕ) and then the radial component obeys modified Bessel's differential equation of order m, determined by the periodicity m of the angular component. One can distinguish two different cases depending on the boundary conditions at infinity (or, alternatively, at $r = 0$). If the solution is supposed to vanish at infinity we choose the "outer" solution denoted as $u_>$. If the solution is supposed to be regular at the origin we choose the "inner" solution $u_<$. Summing up the contributions from all m one obtains

$$u_>(r, \phi) = A_0 K_0(qr) + \sum_{m=1}^{\infty} A_m K_m(qr) \cos(m\phi - m\phi_{>,m}) \qquad (2.22)$$

$$u_<(r, \phi) = B_0 I_0(qr) + \sum_{m=1}^{\infty} B_m I_m(qr) \cos(m\phi - m\phi_{<,m}), \qquad (2.23)$$

where A_0, A_m, B_0, B_m, $\phi_{>,m}$ and $\phi_{<,m}$ are integration constants depending on the specific boundary conditions. I_α and K_α with $\alpha = 0, 1, \ldots$ are modified Bessel functions of the first and the second kind, respectively. The outer solutions of all orders behave at large distances according to

$$K_\alpha(x) \approx \sqrt{\frac{\pi}{2x}} e^{-x}, \quad x \gg 1 \qquad (2.24)$$

while the short distance asymptotics read

$$K_0(x) \approx \ln\left(\frac{2}{x}\right) - \gamma_e, \qquad (2.25)$$

$$K_m(x) \approx 2^{m-1}(m-1)! x^{-m}, \quad x \ll 1, \quad m = 1, 2, \ldots, \qquad (2.26)$$

where γ_e is the Euler-Mascheroni constant. The force balance analogical to Eq. (2.20) can be obtained by integrating Eq. (2.21) over A and multiplying by γ,

$$-\gamma \oint_{\partial A} dl_\parallel \, \boldsymbol{n} \cdot \nabla_\parallel u = -\Delta \rho g \int_A d^2\boldsymbol{x} \, u \qquad (2.27)$$

where the rhs represents minus the weight of the fluid displaced with respect to the reference flat configuration.

Capillary equation at micrometer length scales

When we consider the interface deformations u on a length scale much smaller than the capillary length λ_c, the term $q^2 u$ in Eq. (2.21) can be neglected and the governing equation resembles the two-dimensional Laplace equation. It reads

$$\nabla_\parallel^2 u = 0. \qquad (2.28)$$

The only axisymmetric solution $\bar{u}$ can be written as

$$\bar{u}(r) = A_0 \ln\left(\frac{r}{\zeta}\right) \qquad (2.29)$$

2.2. WETTING

and it is divergent both at small and large distances r. Therefore the constant ζ plays the role of a large-distance cut-off or a small-distance cut-off depending on the specific boundary conditions. The general solution can be expressed in terms of u_0 and the remaining non-axisymmetric inner and outer solutions, which read

$$u_>(r,\phi) = \sum_{m=1}^{\infty} A_m r^{-m} \cos(m\phi - m\phi_{>,m}), \qquad (2.30)$$

$$u_<(r,\phi) = \sum_{m=1}^{\infty} B_m r^{m} \cos(m\phi - m\phi_{<,m}), \qquad (2.31)$$

with the same notation as in Eqs. (2.22) and (2.23). For a given boundary condition at $r = a$ the inner and outer solutions can be joined continuously, i.e., such that their values at $r = a$ are equal. The condition

$$u_>(a,\phi) = u_<(a,\phi) \qquad (2.32)$$

determines B_m as functions of A_m.

2.2 Wetting

When a drop of water is placed on very clean glass it spreads into a thin liquid film, whereas the same drop placed on a sheet of plastic maintains its spherical shape. The different wetting behavior of substrates can be described by so-called spreading parameter S^*, which measures the difference in surface energy per unit area between a dry and a wet substrate (de Gennes et al. 2004),

$$S^* = \gamma_{sg} - \gamma_{sl} - \gamma. \qquad (2.33)$$

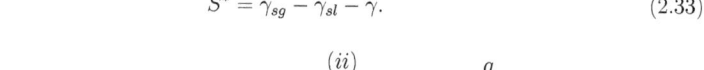
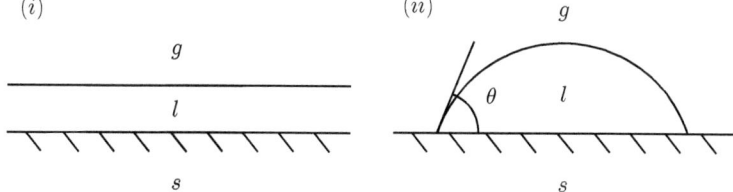

Figure 2.4: The two different physical situations corresponding to (i) complete wetting and (ii) partial wetting.

One can distinguish two regimes (see Fig. 2.4):

(i) $S^* \geq 0$, which corresponds to *complete wetting*, when the liquid lowers its surface energy by spreading over the substrate. This situation corresponds to a vanishing contact angle $\theta = 0$.

(ii) $S^* < 0$, which corresponds to *partial wetting*, when the liquid does not spread but forms a spherical cap with the contact angle θ given by Young's law in Eq. (2.3). As a consequence the spreading parameter S^* can be expressed by θ,

$$S^* = \gamma(\cos\theta - 1). \tag{2.34}$$

We note that the above equation makes sense only if $S^* < 0$.

2.2.1 Wetting energy of a particle at the interface

As mentioned in the Introduction, micron-sized particles can get irreversibly trapped at a liquid-gas interface (or other fluid-fluid interface). This mechanism can be understood in terms of the surface free energy. Consider a smooth solid spherical particle of radius a centered at the height $h < a$ over a flat liquid-gas interface characterized by surface tension γ. The equilibrium position of the particle is determined by the minimum of the sum F of the changes (with respect to a reference configuration which will be specified later) of the surface energies of the particle-liquid contact area S_{pl} characterized by surface tension γ_{pl}, particle-gas contact area S_{pg} characterized by surface tension γ_{pg}, and of the liquid-gas interface S_{lg}:

$$F = \gamma_{pl}\Delta S_{pl} + \gamma_{pg}\Delta S_{pg} + \gamma\Delta S_{lg}. \tag{2.35}$$

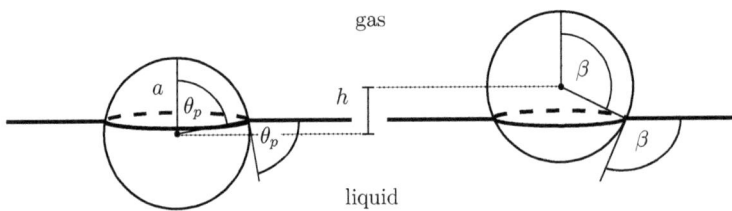

Figure 2.5: Sketches of the configuration of the particle at a flat, undeformable interface. The left panel corresponds to the reference equilibrium configuration $\beta = \theta_p$, and the right panel to a configuration with an arbitrary β.

Due to $\Delta S_{pg} = -\Delta S_{pl}$, after applying Young's law (Eq. (2.3)), we obtain

$$F = \gamma(-\cos\theta_p \Delta S_{pl} + \Delta S_{lg}). \tag{2.36}$$

The contact areas are functions of the polar angle β parameterizing the position of the three-phase contact line at the particle, see Fig. 2.5. Taking as a reference the configuration with $\beta = \theta_p$, one has

$$\Delta S_{pl} = 2\pi a^2(\cos\beta - \cos\theta_p), \tag{2.37}$$
$$\Delta S_{lg} = -\pi a^2(\sin^2\beta - \sin^2\theta_p). \tag{2.38}$$

Thus the total surface free energy in Eq. (2.36) can be expressed as

$$F = \gamma\pi a^2(\cos\beta - \cos\theta_p)^2 = \gamma\pi h^2, \tag{2.39}$$

2.2. WETTING

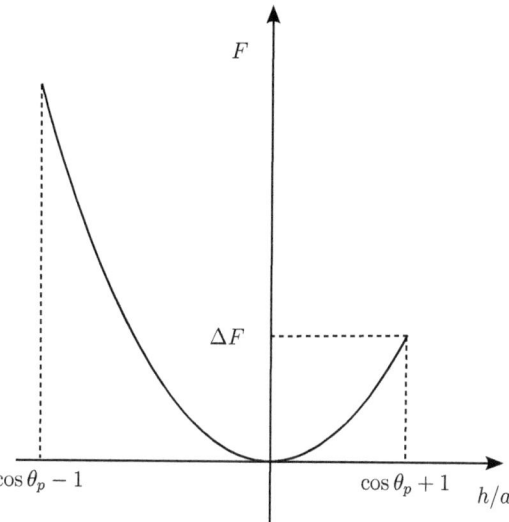

Figure 2.6: The free energy (in arbitrary units) of a particle at a flat undeformable interface, being the sum of the surface free energies Eq. (2.39), as a function of the height h of the particle center above the interface.

where the height $h = a(\cos\beta - \cos\theta_p)$ is measured with respect to the reference configuration. The free energy barrier ΔF for detachment of the particle from the interface is determined by the smaller of the two extremal values of F at $h = a(1 - \cos\theta_p)$ and at $h = a(1 + \cos\theta_p)$, see Fig. 2.6.

2.2.2 Wetting and long-range forces

The macroscopic description of the wetting phenomena in terms of the surface tension is sufficient only in the situations when the thickness of a film or the radius of a droplet exceeds the range of intermolecular forces. However, in the case of very thin films or microscopic droplets the effects of the finite range of the intermolecular forces must be taken into account (de Gennes et al. 2004). The corresponding microscopic approach should recover the macroscopic limit of a thick film, with the surface energy per unit area equal to $\gamma_{sl} + \gamma$, and of a dry substrate, characterized by the surface energy γ_{sg}. This suggests writing the energy per unit area of a film, as a function of the film thickness l, in the form $\gamma_{sl} + \gamma + W(l)$, where the effective interface potential $W(l)$ must obey the condition $W(0) = \gamma_{sg} - \gamma_{sl} - \gamma = S^*$.

In principle, the behavior of the effective interface potential in the limit of small l, governed by the spreading coefficient S^*, can be considered independent of the asymptotic behavior $W(l)$ for l exceeding the molecular length scale. In the systems with van der Waals intermolecular interactions, vanishing at large separations like r^{-6}, the

latter is of the form
$$W(l) = \frac{A_H}{12\pi l^2} \qquad (2.40)$$
where A_H is known as the Hamaker constant, which, depending on the material properties and thermodynamic parameters, can be either positive or negative. The interplay between S^* and the shape of the effective interface potential governed by A_H can lead to various possible wetting scenarios. Here, we pick up a few simple generic cases, whereas for a more detailed analysis we refer to Brochard-Wyart *et al.* (1991). First, we note that an equilibrium film thickness corresponds to a minimum of $W(l)$. If there are more than one minima, the films of a different thickness can coexist in the same sample, which means that liquid forms a droplet. Three generic cases are possible:

(*i*) Complete wetting. Assuming $S^* > 0$ and $A_H > 0$ the minimum of $W(l)$ corresponds to a solid wetted by a macroscopically thick film ($l \to \infty$).

(*ii*) Partial wetting. For $S^* < 0$ and $A_H > 0$ there are two coexisting minima at $l = 0$ and $l \to \infty$, which physically correspond to a sessile droplet surrounded by a dry substrate. In the case $S^* < 0$ and $A_H < 0$ the maximum at $l \to \infty$ can also be regarded as corresponding to a metastable equilibrium. The resulting expression for the contact angle is the same as in the macroscopic approach (see Eq. (2.34)),
$$\cos\theta = 1 + \frac{S^*}{\gamma}. \qquad (2.41)$$

(*iii*) Pseudo-partial wetting. In the case $S^* > 0$ and $A_H < 0$ a film of finite thickness l_0 given by the condition $dW(l)/dl = 0$ coexists with a macroscopically thick film ($l \to \infty$), which corresponds to a macroscopic droplet surrounded by a film of molecular thickness. The effective surface tension of the film $\gamma_{eff} = \gamma + W(l_0)$ must be balanced by the lateral component $\gamma\cos\theta$ of the surface tension of the liquid wedge, from which the macroscopic contact angle reads
$$\cos\theta = 1 + \frac{W(l_0)}{\gamma}. \qquad (2.42)$$

One should note, that when the amount of liquid is insufficient to cover the substrate with a film of thickness l_0 the droplet would spread completely (which justifies the notion of pseudo-partial wetting).

2.2.3 Microscopic approach: density functional theory

Microscopic theory of fluids is nowadays a well established theory on which many monographs and review articles have been published (see, e.g., Evans 1979; Tarazona *et al.* 2008). By no means do we want to present an overview of the theory in this short Subsection, but only to highlight certain issues possibly relevant for the problem of particles at interfaces.

2.2. WETTING

The microscopic density of molecules $\rho(\mathbf{r})$, particularly in presence of an external field due to interactions with a solid substrate, can be addressed in the framework of the density functional theory (DFT) (Evans 1979). Within mean-field approach (neglecting correlations) the theory states that the equilibrium density of a fluid phase exposed to an effective substrate potential $V_s(\mathbf{r})$ minimizes the grand canonical density functional in the form

$$\Omega[\{\rho(\mathbf{r})\}, T, \mu] = \int d^3r\, f_{HS}(\rho(\mathbf{r}), T) + \frac{1}{2}\int d^3r \int d^3r'\, w_f(|\mathbf{r}-\mathbf{r}'|)\rho(\mathbf{r})\rho(\mathbf{r}') \\ + \int d^3r\, (\rho_s V_s(\mathbf{r}) - \mu)\rho(\mathbf{r}), \quad (2.43)$$

where T is the temperature, μ is the chemical potential and ρ_s is the number density of the molecules in the substrate; the integration domain is the whole region accessible to the fluid and $w_f(r)$ describes the long-range attractive part of the fluid-fluid microscopic interaction potential. The short-ranged molecular repulsion, according to the Weeks-Chandler-Andersen (WCA) (Weeks et al. 1971) approximation, is incorporated into the effective hard-sphere free energy density $f_{HS}(\rho, T)$. Consider, for example, the Lennard-Jones pair-potential in the form

$$\phi(r) = 4\epsilon_f \left[\left(\frac{\sigma}{r}\right)^{12} - \left(\frac{\sigma}{r}\right)^6 \right] \quad (2.44)$$

where σ plays the role of the molecular length scale, whereas ϵ_f is the interaction strength. One can perform the decomposition $\phi = \phi_{rep} + \phi_{attr}$, with

$$\phi_{rep} = \begin{cases} \phi(r) + \epsilon_f, & r < 2^{1/6}\sigma, \\ 0, & r > 2^{1/6}\sigma \end{cases} \quad (2.45)$$

and

$$\phi_{attr} = \begin{cases} -\epsilon_f, & r < 2^{1/6}\sigma, \\ \phi(r), & r > 2^{1/6}\sigma, \end{cases} \quad (2.46)$$

where the repulsive part ϕ_{rep} gives rise to an effective, temperature-dependent, hard-sphere diameter $d(T)$. $f_{HS}(\rho, T)$ can then be determined by means of, for example, the Carnahan-Starling approximation (Carnahan & Starling 1969). For analytical treatment it is often convenient to approximate the long-range attractive part ϕ_{attr} by a smooth analytic function. For example $w_f(r)$ can be taken in the form

$$w_f(r) = W_f \frac{\sigma^6}{(\sigma^2 + r^2)^3} \quad (2.47)$$

with the amplitude

$$W_f = -\frac{128\sqrt{2}}{9\pi}\epsilon_f, \quad (2.48)$$

chosen such that

$$\int_{\mathbb{R}^3} d^3r\, w(r) = \int_{\mathbb{R}^3} d^3r\, \phi_{attr}(r). \quad (2.49)$$

Throughout the thesis the width of the liquid-gas interface will be neglected. Thus, DFT will only be used at a semi-microscopic level in order to calculate the shape of the interface in neighborhood of arbitrarily shaped substrates as a function of both the thermodynamic parameters (μ, T) and the microscopic parameters (ϵ_f, σ) of the system. Examples of such calculations for various substrate geometries can be found in the works of Tasinkevych and Dietrich (Tasinkevych & Dietrich 2006; 2007). In the next Chapter we will derive the effective interface potential for a spherical substrate representing a colloidal particle and calculate the deformation of the resulting liquid-vapor interface around the particle, assuming the long-range attraction in the form in Eq. (2.47).

Chapter 3

Stability of a particle at a deformable interface

In this Chapter we address in more detail the issue of the free energy of a single spherical particle at a flat interface, introduced in Chapter 2. First, we apply the macroscopic approach in terms of surface tensions and next we investigate a possible influence of long-range intermolecular forces.

First, we note that in the simple model with a flat, undeformable interface (Eq. (2.39)) the local geometric contact angle, equal to the polar angle β parameterizing the position of the contact line (see fig. 2.5), does not fulfill Young's law apart for the configuration corresponding to the minimum of the free energy in Eq. (2.39), i.e., when $\beta = \theta_p$. In the remaining configurations with $\beta \neq \theta_p$, the interface must get deformed in order to meet the surface of the particle at the given contact angle θ_p (see Fig. 3.1). In this Chapter we provide exact expressions (Eqs. (3.11)-(3.13)) for the free energy $\tilde{F}(h)$ of a single particle at a deformable, finite interface. Those exact results are qualitatively very similar to the results obtained by Gilet & Bush (2009) for a droplet bouncing from a fluid-fluid interface (upon neglecting the deformations of the droplet those results corresponds to a particle with the contact angle $\theta_p = \pi$). The exact solution of the full non-linear problem not only serve as a test for approximate methods applied in the following work, but also reveals the existence of metastable branches of the free energy. Moreover, we demonstrate that those branches can be almost perfectly reproduced by a linear theory for small deformations, renormalized in order to match the asymptotic form of the exact solution far away from the particle (Eq. (3.27)). Finally, we derive an effective potential $\Delta\Omega(h)$ (Eq. (3.62)) starting from the grand canonical density functional $\Omega[\{\rho(\boldsymbol{r})\}]$ for the fluid surrounding the particle and applying the sharp-kink approximation for the density profile $\rho(\boldsymbol{r})$ at the interface. Using an effective interface potential for a spherical substrate (particle), given by a scaling function $t_p(r, a, \sigma)$ of the distance r from the particle center and the particle radius a (Eq. (3.58)) calculated by using the pair-potential in the form in Eq. (2.47), we minimize the functional $\Omega[\{l(\boldsymbol{x})\}]$ with respect to the interface profile $l(\boldsymbol{x})$ numerically. The results, after being normalized by the surface tension expressed as a function of the microscopic system parameters and the temperature, only slightly quantitatively deviate from the macroscopic theory. Thus, we conclude that the macroscopic approach is fully sufficient for the purpose of calculating capillary forces in the case of micron-sized particles.

3.1 Free energy functional

Mechanical equilibrium demands a mechanism immobilizing the interface at a certain distance from the particle. In the case when the gravity is present, the interface is effectively pinned at a distance $r = \lambda_c$ from the particle. For the reason of mathematical simplicity we assume that $g = 0$ and that the interface is pinned at the rim of a cylindrical vessel containing the liquid. However, we note that even in the presence of gravity, the gravitational energy of the fluid displaced due to the deformation of the interface in a close neighborhood of a micrometer-sized particle ($r \ll \lambda_c$) can be neglected with respect to the surface free energy of the liquid-gas interface (see, c.f., Sec. 4.1). Therefore, the assumption $g = 0$ is not necessary as soon as the container size L is much smaller than the capillary length, i.e., $L/\lambda_c \to 0$.

In the case of a deformable interface, the configuration of the system depends not only on the position of the particle but also on the shape of the interface, which, in the case of a spherical particle, due to axial symmetry can be described by the departure $z(r)$ of the interface from the reference flat configuration. The equilibrium shape can be determined by minimizing the following free energy functional,

$$\mathcal{F}[\{z(r)\}, \beta] = -\gamma \cos \theta_p \left(S_{pl}(\beta) - S_{pl,ref}\right) + \gamma \left(S_{lg}[\{z(r)\}, \beta] - S_{lg,ref}\right), \quad (3.1)$$

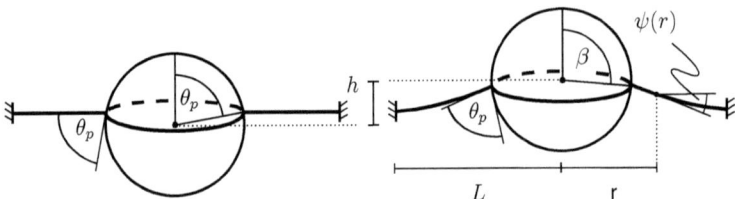

Figure 3.1: Sketches of the configuration of the particle at a deformable interface pinned at a distance L from the particle. The left panel corresponds to the reference configuration $\beta = \theta_p$, and the right panel to a configuration with an arbitrary β. In the latter case β is not uniquely determined by h (see also Fig. 3.4).

In the reference configuration the interface is flat and $h = 0$, i.e., $\beta = \theta_p$ (see Fig. 3.1). The free energy $\hat{F}(\beta)$ under the constraint of fixed β is obtained by minimizing $\mathcal{F}$ with respect to $z(r)$,

$$\hat{F}(\beta) = \min_{\{z(r)\}} \mathcal{F}. \quad (3.2)$$

3.2 The analytic solution of the full Young-Laplace equation

In the following we use the particle radius a as a unit length and γa^2 as a unit energy. The changes in the surface areas can be written as

$$S_{pl} - S_{pl,ref} = 2\pi(\cos\beta - \cos\theta_p), \qquad (3.3)$$

$$S_{lg} - S_{lg,ref} = 2\pi \int_{\sin\beta}^{L} dr\, r\sqrt{1+\dot{z}^2} - \pi(L^2 - \sin^2\theta_p), \qquad (3.4)$$

where $\dot{z} = dz/dr$. Minimizing the functional in Eq. (3.1) with respect to $z(r)$ yields the Young-Laplace equation (Eq. (2.10)) in the form (see, e.g., Langbein 2002),

$$\frac{1}{r}\frac{d}{dr} r \frac{\dot{z}}{\sqrt{1+\dot{z}^2}} = 0. \qquad (3.5)$$

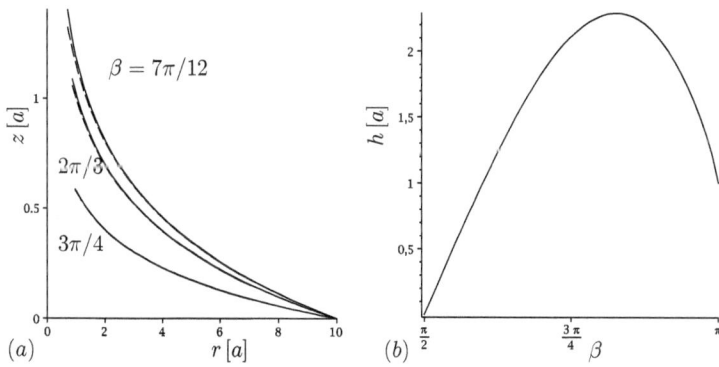

Figure 3.2: (a) Exact interface profiles $z(r)$ (Eq. (3.11)) for $L = 10$, $\theta_p = \pi/2$ and angular positions $\beta = 2\pi/3, 3\pi/4, 5\pi/6$ of the contact line (solid lines) together with the results of the renormalized linear theory $\bar{u}_{ren}(r)$ (Eq. (3.27), dashed lines). (b) Immersion of the particle as a function of β for $L = 10$. The length scale is set by the particle radius.

where $\lambda = 0$ (no volume constraint). The boundary conditions at the particle follow from Young's law, which can be written as

$$\psi(r = \sin\beta) = \beta - \theta_p, \qquad (3.6)$$

where $\psi(r)$ is the angle between the r-axis and the tangent to the profile $z(r)$ defined by

$$\sin\psi(r) := -\frac{\dot{z}}{\sqrt{1+\dot{z}^2}}. \qquad (3.7)$$

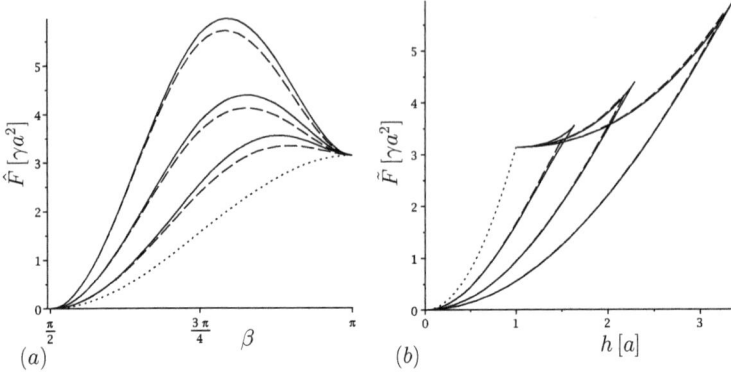

Figure 3.3: The free energy (a) $\hat{F}(\beta)$ and (b) $\tilde{F}(h)$ for various system sizes $L = 2$, 10 and 100, from bottom to top in (a) and from left to right in (b), for $\theta_p = \pi/2$. The dotted line corresponds to the model with an undeformable flat interface (Eq. (2.39)), the solid lines are the exact solution given by Eqs. (3.11)-(3.13) for a deformable interface, and the dashed lines correspond to the renormalized linear theory in Eqs. (3.27)-(3.29).

The first integral of Eq. (3.5) reads

$$\frac{\dot{z}}{\sqrt{1+\dot{z}^2}} = \frac{c}{r}, \tag{3.8}$$

where c is an integration constant, which can be determined from the boundary condition in Eq. (3.6) and by using Eq. (3.7) as

$$c = c(\beta) = \sin\beta \sin(\theta_p - \beta). \tag{3.9}$$

The equilibrium profile $z(r)$ can be obtained by integrating Eq. (3.8) and imposing the second boundary condition

$$z(L) = 0, \tag{3.10}$$

which finally yields

$$z(r) = c\left[\ln(L - \sqrt{L^2 - c^2}) - \ln(r - \sqrt{r^2 - c^2})\right]. \tag{3.11}$$

We note that the immersion h is not an independent variable and should be separately evaluated as a function of the equilibrium interface profile $z(r)$. The latter is uniquely determined by the position of the contact line at the particle expressed in terms of β,

$$h(\beta) = z(r = \sin\beta) - \cos\beta + \cos\theta_p. \tag{3.12}$$

The free energy as a function of β can be obtained by performing the integral in Eq. (3.4) with $z(r)$ given by Eq. (3.11), which yields

$$\hat{F}(\beta) = -2\pi\cos(\theta_p)(1 + \cos\beta) + \pi\left[c(\beta)^2\ln(r + \sqrt{r^2 - c(\beta)^2}) + r\sqrt{r^2 - c(\beta)^2}\right]\Big|_{r=\sin\beta}^{r=L}. \tag{3.13}$$

3.3. MEAN-FORCE ACTING ON THE PARTICLE

Finally, the free energy $\tilde{F}(h) := \hat{F}(\beta)$ is implicitly given by Eqs. (3.11)-(3.13).

The non-monotonic behavior of $h(\beta)$, depicted in Fig. 3.2(b), means that $\beta(h)$ is not uniquely determined for $h > 1$, which leads to the emergence of two branches of $\tilde{F}(h)$ (see Fig. 3.3). Once the system is prepared in a configuration at the metastable branch (the one corresponding to the higher energy) for $h > 1$ and the constraint of fixed h is lifted, the particle slides down the branch by shrinking the contact line and decreasing h, ending up in a state in which it is fully immersed in the gas phase ($h = 1$). As a consequence, a given value of h can be realized by two different solutions, one on the stable and one on the metastable branch. The two solutions are characterized by two different angles β_1 and β_2 (see Fig. 3.4). Actually, there is also an unstable branch being a continuation of the stable branch beyond an inflection point $\partial^2 \tilde{F}/\partial h^2 = -\partial \tilde{f}/\partial h = 0$. However, we postpone a detailed discussion of all the branches of the free energy to Appendix E, in which we consider the case of a particle at the surface of a sessile droplet (which is actually more general due to the non-vanishing mean curvature).

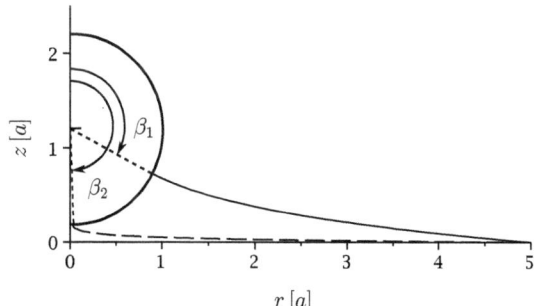

Figure 3.4: Two different interface profiles (Eq. (3.11)) for the fixed immersion $h = 1.2$ characterized by the angles $\beta_1 = 2.05$ and $\beta_2 = 3.11$ (solid and long-dashed lines, respectively), being the two solutions of $h(\beta) = 1.2$ (Eq. (3.12)). The profiles for β_1 and β_2 correspond to the stable and the metastable branches of the free energy (Fig. 3.3(b)), respectively. If the constraint of fixed h is released then the particle moves towards either the globally stable reference configuration ($h = 0$, $\beta = \theta_p = \pi/2$) or towards the metastable configuration ($h = 1$, $\beta = \pi$), respectively.

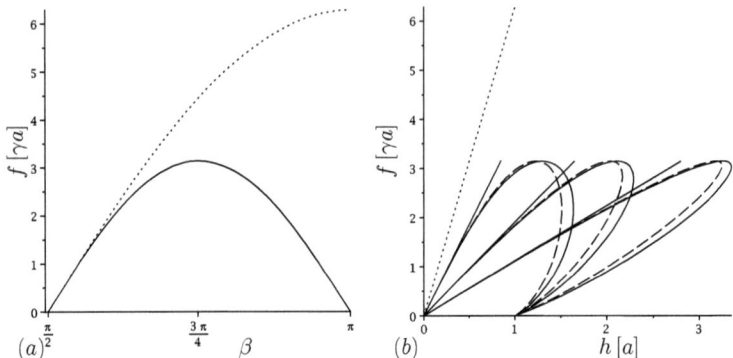

Figure 3.5: The external force f needed to counterbalance the mean-force $\tilde{f}$ ($f = -\tilde{f}$) as a function of β (i.e., $f = -\hat{f}(\beta)$) in (a) and as a function of h in (b), see Eqs. (3.19), (3.17), (3.12) and (3.18). In (b) one can distinguish two branches (in the case of the exact solution as well as in the case of the renormalized linear theory): $\tilde{f} = \tilde{f}_1(h)$ starting at $h = 0$ and finishing at $h = h_{max}$ and $\tilde{f} = \tilde{f}_2(h)$ starting at $h = h_{max}$ and finishing at $h = 1$. The meaning of the lines corresponds to Fig. 3.3.

3.3 Mean-force acting on the particle

The mean-force $\tilde{f}(h)$ acting on the particle is formally given by minus the derivative of the free energy $\tilde{F}(h)$ with respect to h, which can be calculated using the chain rule

$$\tilde{f}(h) = -\frac{\partial \tilde{F}}{\partial h} = -\frac{\partial \hat{F}}{\partial \beta}\left(\frac{\partial h}{\partial \beta}\right)^{-1} \quad (3.14)$$

However, the rhs is ill defined for $\partial h/\partial \beta = 0$, i.e., at the extremal value of h. Indeed, at those points, as can be inferred from Fig. 3.3(b), the free energy has a maximum such that $\partial \hat{F}/\partial \beta = 0$ (Fig. 3.3(a)). In fact, there are two branches $\tilde{f}_1(h)$ and $\tilde{f}_2(h)$, where $\tilde{f}_1(h) > \tilde{f}_2(h)$ (see Fig. 3.5(b)), such that $\tilde{f}(h) = \tilde{f}_1(h)$ for the configurations with $\beta \in [\pi/2, \beta_{max}]$ and $\tilde{f}(h) = \tilde{f}_2(h)$ for $\beta \in [\beta_{max}, \pi]$, where β_{max} indicates the extremum (maximum) of $h(\beta)$ (see Fig. 3.2(b)). As a consequence, for example, in Fig. 3.4 the two configurations corresponding to a given h correspond to different forces. Mechanically, the force acting *on the particle* can be calculated as minus the force acting *on the interface* at the contact line C (there is no contribution from the pressure because $\lambda = 0$), the latter being given by the lhs of Eq. (2.18), so that

$$\tilde{\boldsymbol{f}}_{cap} = -\gamma \oint_C dl\, \boldsymbol{e}_t, \quad (3.15)$$

where the unit vector $\boldsymbol{e}_t$ has been defined in Eq. (2.12) (such that $-\boldsymbol{e}_t$ points outside the particle). Due to axial symmetry, only the vertical component of this force does

3.3. MEAN-FORCE ACTING ON THE PARTICLE

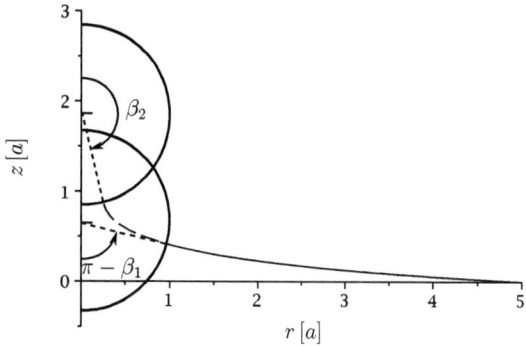

Figure 3.6: Two different interface profiles (Eq. (3.11)) for $\theta_p = \pi/2$ and for the fixed value of the external force $f = \pi/2$. The corresponding two different angular positions of the contact line, $\beta_1 = 7\pi/12$ and $\beta_2 = 11\pi/12$ being the two solutions of $f = -\hat{f}(\beta) = -\pi \sin(2\beta)$ are depicted by the short-dashed line. In the latter case the interface is a continuation (depicted by the long dashed line) of the same analytic form as the common solution (the solid line).

not vanish, i.e., $\tilde{\boldsymbol{f}}_{cap} = \tilde{f}_{cap}\boldsymbol{e}_z$. Taking into consideration Young's law in the form of Eq. (3.6), it can be written as

$$\tilde{f}_{cap} = -2\pi \sin\beta \sin\psi(r = \sin\beta) = 2\pi \sin\beta \sin(\theta_p - \beta) = 2\pi c(\beta) \equiv \hat{f}_{cap}(\beta), \quad (3.16)$$

where we have introduced the notation $\hat{f}_{cap}$ for the capillary force as a function of β. It can be shown by an explicit calculation that $\tilde{f}$ defined in Eq. (3.14) is numerically equal to $\tilde{f}_{cap}$. However, this would not be the case for the interface with a non-vanishing mean-curvature, see Appendix E, when the internal pressure of the liquid λ would not vanish. Nevertheless, in this Section, for simplicity, we omit the index cap.

According to Eq. (3.16) the capillary force depends only on β and not on the system size L. For $\theta_p = \pi/2$ one obtains

$$\hat{f} = \pi \sin(2\beta). \quad (3.17)$$

Eq. (3.17) implies that in this case the absolute value of the force is bounded by π, which is precisely twice smaller then in the unrealistic model with an undeformable flat interface (see Fig. 3.5). In the latter case the force $\tilde{f}_{flat}$, as calculated from Eq. (3.14) with $\tilde{F} = F = \pi\gamma h^2$ (Eq. (2.39)), reads

$$\tilde{f}_{flat} = -\frac{dF}{dh} = -2\pi\gamma h = -2\pi\gamma(\cos\theta_p - \cos\beta). \quad (3.18)$$

42 CHAPTER 3. STABILITY OF A PARTICLE AT A DEFORMABLE INTERFACE

The capillary force, as defined in Eq. (3.15) vanishes for a perfectly flat interface. However, as soon as $\beta \neq \theta_p$, the contact angle is different from θ_p which violates Young's law. As a consequence, there is a thermodynamic "wetting" force $\gamma(\cos\beta - \cos\theta_p)$ pulling the particle towards the reference configuration in which Young's law holds. Upon releasing the constraint of fixed shape of the interface the free energy $F = \gamma h^2$ is lowered towards $\tilde{F}$ and the force acting on the particle is also smaller in the latter case, i.e., $|\tilde{f}| = |-\partial \tilde{F}/\partial h| < |-dF/dh| = |\tilde{f}_{flat}|$ (see Fig. 3.5).

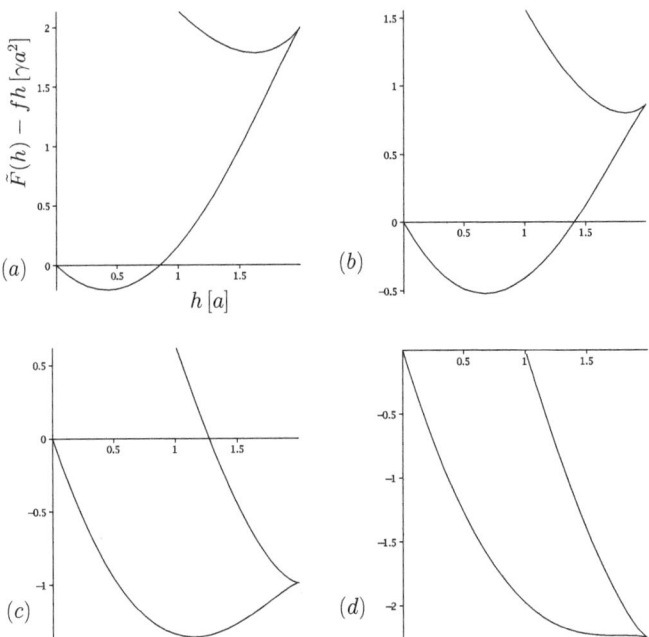

Figure 3.7: (a), (b), (c), and (d): the effective potential $\tilde{F}(h) - fh$ for a particle subjected to an external force $f = 1$, $\pi/2$, 2.5, and π, respectively, for $L = 5$ and $\theta_p = \pi/2$.

The external force f, which must counterbalance the capillary force $\tilde{f}(h)$ in order to fix the particle at a given height h, equals

$$f = -\tilde{f}(h), \qquad (3.19)$$

The interface profiles $z(r)$ (Eq. (3.11)) for $h = h_1$ and $h = h_2$ where $h_{1,2}$ are two distinct solutions of Eq. (3.19) has been plotted in Fig. 3.6 in the case $\theta_p = \pi/2$. According to Eqs. (3.11) and (3.16) $z(r)$ is uniquely determined by the external force, so that those interface profiles are identical for $r > \sin\beta_1$, whereas for $\sin\beta_2 < r < \sin\beta_1$ the first solution is undefined and the second one is a smooth continuation of the common solution. The configuration h_1 corresponds to the stable branch of the free energy whereas h_2 to the metastable one, so that $\tilde{F}(h_1) < \tilde{F}(h_2)$.

We emphasize that the configurations corresponding to $h \neq 0$ can only be realized by applying an external force as given by Eq. (3.19). In such a case the particle experiences an effective potential $\tilde{F}(h) - fh$, whose global or local (non-global) minima correspond to the equilibrium stable or metastable configurations, respectively. As already mentioned, the capillary force, counterbalancing the external force, cannot exceed certain critical values, which determine the range of stability. For $\theta_p = \pi/2$ those values equal $\pm \pi$ (see Eq. (3.17)) such that for $|f| > \pi$ the potential $\tilde{F}(h) - fh$ does not exhibit a minimum and the particle detaches from the interface. For $|f| < \pi$ there is at least one minimum and the particle has an equilibrium configuration at the interface (see Figs. 3.7(a),(b),(c)). The case $|f| = \pi$ separates the two regimes and in this case the effective potential $\tilde{F}(h) - fh$ has an inflection point (see Fig. 3.7(d)). The particle can also get trapped in metastable configurations for $|f| < |f_{meta}| < \pi$, where $f_{meta} = \lim'_{h \to h_{max}} \partial \tilde{F}/\partial h = -\lim'_{h \to h_{max}} \tilde{f}(h)$, where $\lim'$ means that the limit is taken on the metastable branch of $\tilde{F}$ (see Figs. 3.7(a),(b)). By comparing the plots in Fig. 3.5(b) for various L one can see that, with increasing L, $|f_{meta}|$ increases towards the maximal value of the force equal to π.

3.4 Linear theory and renormalization

Under the condition of small deformations $u(r) = z(r)$, such that $|du/dr| \ll 1$, the Young-Laplace equation (3.5) can be linearized and reads

$$\frac{1}{r}\frac{d}{dr} r \frac{du}{dr} = 0. \tag{3.20}$$

We obtain a set of solutions analogical to Eqs. (3.11)-(3.13):

$$u(r) = b \ln\left(\frac{r}{L}\right), \tag{3.21}$$

$$h_{lin}(\beta) = u(\sin\beta) - \cos(\beta), \tag{3.22}$$

$$\hat{F}_{lin}(\beta) = -2\pi \cos\theta_p (1 + \cos\beta) - \pi \sin^2\beta + \pi b^2 \ln\left(\frac{L}{\sin\beta}\right), \tag{3.23}$$

where h_{lin} and $\hat{F}_{lin}$ are the immersion and the free energy in the linear theory, respectively, and b is an integration constant which can be determined from the boundary condition at the particle. From the first integral of Eq. (3.20) in the form

$$\tan\psi = \frac{b}{r}, \tag{3.24}$$

evaluated at $r = \sin\beta$, and using Eq. (3.6) we obtain

$$b = b(\beta) = \cos\beta. \tag{3.25}$$

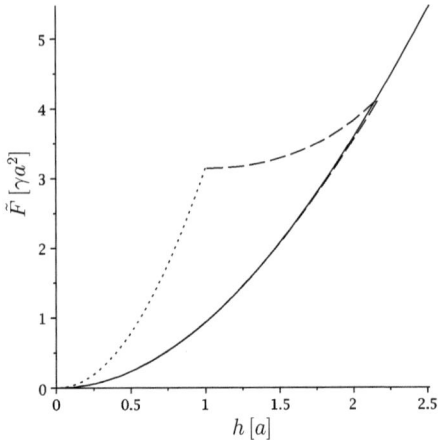

Figure 3.8: Comparison of the free energies calculated within the linear theory as $\tilde{F}_{lin}(h) := \hat{F}_{lin}(\beta_{lin}(h))$ (solid line, see Eq. (3.23)) and the renormalized linear theory $\tilde{F}_{ren}(h) := \hat{F}_{ren}(\beta_{ren}(h))$ (dashed line, see Eq. (3.29)), where $\beta_{lin}(h)$ and $\beta_{ren}(h)$ are given implicitly by Eqs. (3.22) and (3.28). Only $\beta_{ren}(h)$ has two branches, therefore the metastable branch appears only in the renormalized theory. The dotted line is a parabolic free energy profile (Eq. (2.39)).

The above expressions are expected to hold only for the values of β close to θ_p such that the interface is almost flat everywhere. Otherwise, the solution in Eq. (3.21) does not apply close to the particle, where the deformation is varying rapidly. However, even in this case, if the system size L is large enough, the interface becomes flat far away from the particle and Eqs. (3.5) and (3.20) are equivalent. Therefore one can treat the particle with surrounding interface as an effective particle. This renormalization corresponds to replacing the amplitude b by an effective amplitude b_{eff}, such that the two solutions (3.11) and (3.21) match asymptotically. Because $z(r)$ in Eq. (3.11) behaves for large distances like $c \ln(r/L)$, we simply take

$$b_{\text{eff}} \equiv c = \frac{\hat{f}(\beta)}{2\pi} = -\frac{f}{2\pi}, \qquad (3.26)$$

where we have used Eqs. (3.16) and (3.19). The renormalized linear equilibrium deformation (see Fig. 3.2(a)) reads

$$u_{ren}(r) = \frac{f}{2\pi} \ln\left(\frac{L}{r}\right), \qquad (3.27)$$

which recovers the well known expression for the deformation around a heavy particle (Oettel et al. 2005b), with L playing the role of the capillary length λ_c. After the renormalization the displacement of the particle and the free energy read

$$h_{ren}(\beta) = u_{ren}(\sin\beta) - \cos(\beta), \qquad (3.28)$$

3.5. DEPENDENCE ON THE SYSTEM SIZE

and

$$\hat{F}_{ren}(\beta) = -2\pi \cos\theta_p(1 + \cos\beta) - \pi \sin^2\beta + \frac{\hat{f}(\beta)^2}{4\pi} \ln\left(\frac{L}{\sin\beta}\right), \quad (3.29)$$

with $\hat{f}(\beta)$ given by Eq. (3.16). We note that only after the renormalization the metastable branches of the free energy are recovered (in the non-renormalized linear theory the free energy $\hat{F}_{lin}(\beta)$ diverges for $\beta \to 0, \pi$, see Eq. (3.23) and Fig. 3.8).

3.5 Dependence on the system size

The free energy $\tilde{F}(h)$ can be approximated around its minimum at $h = 0$ by a harmonic potential with an effective spring constant k

$$\tilde{F}(h) \approx \frac{kh^2}{2} + \ldots \quad (3.30)$$

Differentiating Eq. (3.30) twice with respect to h and changing the independent variable to β we obtain

$$k = k(L) = -\frac{d\hat{f}}{d\beta}\left(\frac{dh}{d\beta}\right)^{-1}\bigg|_{\beta=\theta_p}. \quad (3.31)$$

For β close to θ_p the relation $h(\beta)$ in Eq. (3.12) can be replaced by $h_{ren}(\beta)$ in Eq. (3.28) and we finally obtain

$$k(L) = \frac{2\pi}{\ln L + 1}, \quad (3.32)$$

which is a slowly decreasing function of L. As can be checked, this result is actually general, in the sense that it is not only limited to the linear theory, which is no surprise because in the limit $h \to 0$ the linear theory is actually exact (for $h = 0$ the deformation vanishes). While the spring constant k decreases with the system size, the free energy for a given β increases, as can be seen from the asymptotic form

$$\hat{F}(\beta) \xrightarrow[L\to\infty]{} \frac{\hat{f}(\beta)^2}{4\pi} \ln L + O(1). \quad (3.33)$$

which is valid for $\beta \in [0, \pi]$. The subleading term $O(1)$ is responsible for the asymmetry of $\hat{F}(\beta)$ with respect to $\beta = 3\pi/4$ for $\beta \in [\pi/2, \pi]$, see Fig. 3.3(a). One can see that upon increasing L the maximum of $\hat{F}$ moves towards $\beta = 3\pi/4$, but the offset $\hat{F}(\beta = \pi) - \hat{F}(\beta = \pi/2) = \pi$ is independent on L. Therefore, for a given L and for β sufficiently close to the value $\beta = \pi$, i.e., close to the minimum of the metastable branch, the term $O(1)$ dominates. Thus, Eq. (3.33) does not grasp the behavior of the free energy close to the minimum of the metastable branch. Nevertheless, apart from the close neighborhood of $\beta = \pi$, for $L \to \infty$ the free energy is determined exclusively by the capillary force, which means that in this limit the wetting energy of the particle (the term $S_{pl} - S_{pl,ref}$ in Eq. (3.1)) is subdominant. Thus, in this regime the free energy is independent on θ_p. Remembering that $f = -\hat{f}(\beta)$ the free energy depends only on the external force. Furthermore, it follows from Eq. (3.33) that the energy barrier between the stable and the metastable branches increases logarithmically with L. More precisely, the maximum of the free energy in the limit $L \to \infty$ reads $\hat{F}_{max} = (\hat{f}(\beta = 3\pi/4))^2 \ln L/(4\pi) + O(1) = (\pi/4)\ln L + O(1)$. Thus, the two branches of the free energy become mutually inaccessible. This can be understood by the fact that, for example, the shapes of the interface are significantly different for two distinct values of β corresponding to a given h (see Fig. 3.4). Therefore, with growing L the difference in the liquid-gas surface area and thus the difference in the free energy between these two states also grows. Finally, we note that the interface profile as given by either the full solution $z(r)$ or by the renormalized linear theory $u_{ren}(r)$ is uniquely determined by the external force (see Eqs. (3.11), (3.16) and (3.27), and Fig. 3.6) independently of L, i.e., not only in the limit $L \to \infty$.

3.6 Influence of long-range intermolecular forces

In this Section we address the problem of a spherical particle at the liquid-vapor interface in the microscopic approach. We want to study how long-range intermolecular forces influence the stability of the macroscopic global equilibrium configuration with the particle at a flat interface. We assume that without the particle the interface is perfectly flat and localized at $z = 0$, which can be achieved by introducing an infinitesimal gravitational field g or a flat substrate underneath the interface (then the interface would be actually a part of a liquid film of thickness l). We assume that either the gravity or long-range interactions with the substrate are responsible only for the localization of the interface, but otherwise their effect can be neglected (this means $g \to 0$ or $l \gg a$).

Our starting point is the grand canonical density functional $\Omega[\{\rho(\bm{r})\}, T, \mu]$ defined in Eq. (2.43) for the fluid around a spherical particle composed of molecules of number density ρ_p and centered at the height h over an initially flat interface:

$$\Omega[h, \{\rho(\bm{r})\}; T, \mu] = \int d^3r\, f_{HS}(\rho(\bm{r}), T) + \frac{1}{2} \int d^3r \int d^3r'\, w_f(|\bm{r} - \bm{r}'|) \rho(\bm{r}) \rho(\bm{r}') \\ + \int d^3r\, (\rho_p V_p(h, \bm{r}) - \mu) \rho(\bm{r}), \quad (3.34)$$

where $V_p(h, \bm{r})$ is the external potential due to the particle, which reads

$$V_p(h, \bm{r}) = \int_{\mathcal{V}_p} d^3r'\, w_p(|\bm{r} - \bm{r}'|) \quad (3.35)$$

where $\mathcal{V}_p$ denotes the spherical domain occupied by the molecules inside the particle. We assume that the fluid-particle pair potential $w_p(r)$ has the same functional form that w_f except for the prefactor $W_p = W_f \epsilon_p / \epsilon_f$, where ϵ_p is the strength of the particle-fluid interaction. We also assume a step-like varying density profile $\rho(\bm{r})$ with a kink at the interface, i.e., for $\bm{r} \in S$, such that $\rho(\bm{r}) = \rho_l$ inside the domain $\mathcal{V}_l$ occupied by liquid and $\rho(\bm{r}) = \rho_g$ inside the domain $\mathcal{V}_g$ occupied by gas, which can be written as

$$\rho(\bm{r}) = \begin{cases} \rho_l & \text{for } \bm{r} \in \mathcal{V}_l, \\ \rho_g & \text{for } \bm{r} \in \mathcal{V}_g, \end{cases} \quad (3.36)$$

3.6. INFLUENCE OF LONG-RANGE INTERMOLECULAR FORCES

and where $S = \partial \mathcal{V}_l \cap \partial \mathcal{V}_g$. Correspondingly, the functional in Eq. (3.34) attains the form

$$\Omega[h, \{\mathcal{V}_l\}; \rho_l, \rho_g, T, \mu] = (f_{HS}(\rho_l, T) - \mu\rho_l)\mathcal{V}_l + (f_{HS}(\rho_g, T) - \mu\rho_g)\mathcal{V}_g$$
$$+ \frac{W_f}{2}\sigma^6 \int_{\mathcal{V}_l+\mathcal{V}_g} d^3r \int_{\mathcal{V}_l+\mathcal{V}_g} d^3r' \frac{\rho(\mathbf{r})\rho(\mathbf{r}')}{(\sigma^2 + |\mathbf{r}-\mathbf{r}'|^2)^3}$$
$$+ W_p\sigma^6 \int_{\mathcal{V}_l+\mathcal{V}_g} d^3r \int_{\mathcal{V}_p} d^3r' \frac{\rho(\mathbf{r})\rho_p}{(\sigma^2 + |\mathbf{r}-\mathbf{r}'|^2)^3}, \quad (3.37)$$

where $\{\mathcal{V}_l\}$ denotes the set of all possible shapes of the liquid domain which fulfill $\mathcal{V}_l \cap \mathcal{V}_p = \emptyset$ and where $\mathcal{V}_g$ is determined as a compliment of $\mathcal{V}_l$ to the whole space accessible to the fluid. Eq. (3.37) incorporates dimensionless integrals of the form

$$I_{\alpha\beta} = \int_{\mathcal{V}_\alpha} d^3r \int_{\mathcal{V}_\beta} d^3r' (\sigma^2 + |\mathbf{r}-\mathbf{r}'|^2)^{-3}, \quad (3.38)$$

where $\alpha, \beta = p, l, g$. Using the divergence theorem the volume integrals can be transformed into surface integrals by introducing a second-rank tensor $\mathbf{T}(\mathbf{r}, \mathbf{r}', \sigma)$:

$$I_{\alpha\beta} = \frac{1}{\sigma^3} \int_{\mathcal{V}_\alpha} d^3r \int_{\mathcal{V}_\beta} d^3r' \nabla'\nabla : \mathbf{T}(\mathbf{r}, \mathbf{r}', \sigma)$$
$$= \frac{1}{\sigma^3} \int_{S_\alpha} d^2S \int_{S_\beta} d^2S' \mathbf{n}_\alpha(\mathbf{r}) \cdot \mathbf{T}(\mathbf{r}, \mathbf{r}', \sigma) \cdot \mathbf{n}_\beta(\mathbf{r}'), \quad (3.39)$$

where $S_{\alpha,\beta}$ denote surfaces enclosing volumes $\mathcal{V}_{\alpha,\beta}$ and $\mathbf{n}_{\alpha,\beta}$ are the corresponding normals directed outwards. The tensor $\mathbf{T}$ can be determined from the following differential equation

$$\nabla'\nabla : \mathbf{T}(\mathbf{r}, \mathbf{r}', \sigma) = \frac{\sigma^3}{(\sigma^2 + |\mathbf{r}-\mathbf{r}'|^2)^3}. \quad (3.40)$$

We make the following ansatz:

$$\mathbf{T}(\mathbf{r}, \mathbf{r}', \sigma) = t(|\mathbf{r}-\mathbf{r}'|, \sigma)\,\mathbf{1}, \quad (3.41)$$

which after inserting into Eq. (3.40) results in a differential equation for the function $t(r, \sigma)$:

$$-\frac{1}{r^2}\frac{d}{dr}r^2\frac{d}{dr}t(r, \sigma) = \frac{\sigma^3}{(r^2 + \sigma^2)^3}, \quad (3.42)$$

with the solution:

$$t(r, \sigma) = \frac{1}{8\sigma}\left[\frac{\sigma}{r}\left(A + \arctan\left(\frac{r}{\sigma}\right)\right) + \frac{\sigma^2}{r^2 + \sigma^2}\right] + B. \quad (3.43)$$

The integration constants A and B must match the expected asymptotics of $t(r, \sigma)$. Without loosing generality we can put $B = 0$, because the additive constant leads to a vanishing contribution to the surface integrals in Eq. (3.39) due to the fact that $\int_S d^2S\,\mathbf{n} = 0$ for an arbitrary closed surface S. The constant A must be chosen more

48 CHAPTER 3. STABILITY OF A PARTICLE AT A DEFORMABLE INTERFACE

carefully. First, we note that in the limit $\sigma \to 0$ the rhs of Eq. (3.40) vanishes everywhere except for $r = r'$, whereas its integral over $\mathbb{R}^3$ remains constant:

$$\int_{\mathbb{R}^3} d^3 r \frac{\sigma^3}{(\sigma^2 + |r - r'|^2)^3} = 4\pi \int_0^\infty dx \frac{x^2}{(1 + x^2)^3} = \frac{\pi^2}{4}. \tag{3.44}$$

Therefore the rhs of Eq. (3.40) in the limiting case $\sigma \to 0$ behaves as $(\pi^2/4)\delta(r - r')$, where $\delta(r - r') = \delta(|r - r'|)/(4\pi |r - r'|^2)$ is the delta function. Thus, one can rewrite Eq. (3.40) as

$$\nabla' \nabla : T(r, r', \sigma) = \frac{\pi \delta(|r - r'|)}{16 |r - r'|^2} + \Delta(|r - r'|, \sigma), \tag{3.45}$$

where the function $\Delta(r, \sigma) := \sigma^3 (\sigma^2 + r^2)^{-3} - (\pi/16 r^2) \delta(r)$ vanishes identically for $\sigma \to 0$. Thus, in this limit the function $t(r, \sigma \to 0) =: t_0(r)$ fulfills Eq. (3.42) with the rhs replaced by $(\pi/16 r^2) \delta(r)$, i.e.,

$$-\frac{1}{r^2} \frac{d}{dr} r^2 \frac{d}{dr} t_0(r) = \frac{\pi \delta(r)}{16 r^2}. \tag{3.46}$$

Multiplying both sides by r^2 and integrating leads to

$$\frac{d}{dr} t_0(r) = -\frac{\pi}{16 r^2}, \tag{3.47}$$

which gives

$$t_0(r) = \frac{\pi}{16 r}. \tag{3.48}$$

According to Eq. (3.43) one has $t_0(r) = t(r, \sigma \to 0) = (A + \pi/2)/(8r)$ which implies that $A = 0$. It is useful to split the function $t(r, \sigma)$ into $t_0(r)$ and the remaining part $\delta t(r, \sigma)$, which reads

$$\delta t(r, \sigma) := t(r, \sigma) - t_0(r) = \frac{1}{8\sigma} \left[\frac{\sigma}{r} \left(\arctan\left(\frac{r}{\sigma}\right) - \frac{\pi}{2} \right) + \frac{\sigma^2}{r^2 + \sigma^2} \right]. \tag{3.49}$$

Then, by using Eq. (3.41), the integral in Eq. (3.38) may be written as

$$\begin{aligned}
I_{\alpha\beta} &= \frac{1}{\sigma^3} \int_{S_\alpha} d^2 S \int_{S_\beta} d^2 S' \, n_\alpha(r) \cdot n_\beta(r') t(|r - r'|, \sigma) \\
&= \frac{1}{\sigma^3} \int_{S_\alpha} d^2 S \int_{S_\beta} d^2 S' \, n_\alpha(r) \cdot n_\beta(r') \left[t_0(|r - r'|) + \delta t(|r - r'|, \sigma) \right] \\
&= \frac{\pi^2}{4 \sigma^3} \int_{\mathcal{V}_\alpha} d^3 r \int_{\mathcal{V}_\beta} d^3 r' \, \delta(r - r') + \frac{1}{\sigma^3} \int_{S_\alpha} d^2 S \int_{S_\beta} d^2 S' \, n_\alpha(r) \cdot n_\beta(r') \delta t(|r - r'|, \sigma) \\
&= \frac{\pi^2}{4 \sigma^3} (\mathcal{V}_\alpha \cap \mathcal{V}_\beta) + \frac{1}{\sigma^3} \int_{S_\alpha} d^2 S \int_{S_\beta} d^2 S' \, n_\alpha(r) \cdot n_\beta(r') \delta t(|r - r'|, \sigma) \\
&:= I_{0,\alpha\beta} + \delta I_{\alpha\beta} \tag{3.50}
\end{aligned}$$

where in the last equality we have separated the contribution $I_{0,\alpha\beta}$, stemming from $t_0(r)$ and scaling with the volume of the common part of the domains $\mathcal{V}_\alpha$ and $\mathcal{V}_\beta$ but

3.6. INFLUENCE OF LONG-RANGE INTERMOLECULAR FORCES

being independent of their shapes, from the remaining contribution $\delta I_{\alpha\beta}$, depending on the shapes of the domains.

The total contribution to the density functional in Eq. (3.37) depending on the volume of liquid can be written as $\Delta\Omega_b = \mathcal{V}_l \times [\omega_b(\rho_l, T, \mu) - \omega_b(\rho_g, T, \mu)]$, where ω_b is the grand canonical potential density of a bulk fluid. Close to the bulk liquid-gas coexistence line $\mu = \mu_0(T)$ on has $\Delta\Omega_b = \mathcal{V}_l \Delta\rho \Delta\mu$, with $\Delta\mu = \mu_0(T) - \mu$ and $\Delta\rho = \rho_l - \rho_g$. In the following we consider only the situation of liquid-gas coexistence, for which the theory yields finite contact angles at flat substrates (see, c.f., Eq. (3.55)). In such a case $\Delta\Omega_b = 0$, whereas the number densities $\rho_l = \rho_l(T, \mu_0(T))$ and $\rho_g = \rho_g(T, \mu_0(T))$ are functions of the temperature only. The grand canonical density functional reads

$$\Omega_{coex}[h, \{\mathcal{V}_l\}; T] = const + \frac{W_f}{2}(\sigma^3 \Delta\rho)^2 \delta I_{ll} + (W_p \rho_p - W_f \rho_g) \sigma^6 \Delta\rho \, \delta I_{pl}. \quad (3.51)$$

To this end, we have not made any assumptions concerning the shape of the particle and in this sense the functional in Eq. (3.51) is general. We can also use it to derive the contact angle at the particle. We replace the particle by a flat substrate composed of the same kind of molecules, i.e., characterized by the same number density ρ_p and the same pair-potential w_p. A thin film of liquid in contact with the substrate forms two flat interfaces of surface area S separated by a film of thickness l. The grand canonical potential, under the condition of liquid-gas coexistence, can be then written as

$$\Omega_{coex}^{flat}(l, T) = S\big(\gamma + W(l)\big), \quad (3.52)$$

where

$$\gamma = \frac{W_f}{2}\sigma^6(\Delta\rho)^2 \frac{1}{\sigma^4}\int_0^{2\pi}d\phi\int_0^\infty dr\, r\, \delta t(r, \sigma) = \frac{16\sqrt{2}(\sigma^3\Delta\rho)^2 \epsilon_f}{9\sigma^2} \quad (3.53)$$

can be identified as the surface tension of a flat liquid-gas interface and

$$W(l) = -\frac{32\sqrt{2}}{9}\sigma^4 \Delta\rho(\epsilon_f \rho_l - \epsilon_p \rho_p)\left(1 + \frac{l}{\sigma}\arctan\left(\frac{l}{\sigma}\right) - \frac{\pi l}{2\sigma}\right) \quad (3.54)$$

as the effective interface potential (Tasinkevych & Dietrich 2007). Thus the macroscopic contact angle at the particle (Eq. (2.42)) can be expressed in terms of microscopic system parameters and number densities as

$$\cos\theta_p = 1 + \frac{W(0)}{\gamma} = 1 - \frac{2(\epsilon_f \rho_l - \epsilon_p \rho_p)}{\epsilon_f \Delta\rho} \quad (3.55)$$

Accordingly, the density functional for the spherical particle of the radius a can be written as

$$\Omega_{coex}^{sph}[h, \{\mathcal{V}_l\}; a, \theta_p] = -\frac{4}{\pi}\gamma\sigma^2 [\delta I_{ll} + (1 + \cos\theta_p)\delta I_{pl}] \quad (3.56)$$

with γ given in Eq. (3.53), $\delta I_{ll} = \delta I_{ll}[\{l(\boldsymbol{x})\}]$ and $\delta I_{pl} = \delta I_{pl}[h, a; \{l(\boldsymbol{x})\}]$. The integral over the spherical surface of the particle in the expression for δI_{pl} can be carried out analytically with the result

$$\delta I_{pl} = \frac{1}{\sigma^3}\int_{S_l} d^2 S\, \boldsymbol{n}_l \cdot \frac{\boldsymbol{r} - h\boldsymbol{e}_z}{|\boldsymbol{r} - h\boldsymbol{e}_z|} t_p(|\boldsymbol{r} - h\boldsymbol{e}_z|, a, \sigma), \quad (3.57)$$

where

$$t_p(r, a, \sigma) = \frac{\pi}{12r^2} \left[r^3 \left(\arctan\left(\frac{r+a}{\sigma}\right) - \arctan\left(\frac{r-a}{\sigma}\right) \right) \right.$$
$$\left. + a^3 \left(\arctan\left(\frac{r+a}{\sigma}\right) + \arctan\left(\frac{r-a}{\sigma}\right) - \pi \right) - 2ar\sigma + \frac{\sigma^3}{2} \log \frac{(r+a)^2 + \sigma^2}{(r-a)^2 + \sigma^2} \right], \tag{3.58}$$

The quantity δI_{ll} contains a double integral over the liquid-gas interface S and therefore it cannot be further simplified. In the *local* approximation one of the integrals can be replaced by an integral over an infinite plane, so that

$$\delta I_{ll} \approx -\frac{\pi}{4\sigma^2} \int_S d^2 S, \tag{3.59}$$

and then the functional in Eq. (3.56) can be written as

$$\Omega^{sph}_{coex,loc}[h, \{\mathcal{V}_l\}; a, \theta_p] =$$
$$= \gamma \int_S d^2 S \left[1 - \frac{4}{\pi}(1 + \cos\theta_p) \, \boldsymbol{n}_l \cdot \frac{\boldsymbol{r} - h\boldsymbol{e}_z}{|\boldsymbol{r} - h\boldsymbol{e}_z|} t_p(|\boldsymbol{r} - h\boldsymbol{e}_z|, a, \sigma) \right]. \tag{3.60}$$

where $\boldsymbol{r} = \boldsymbol{x} + l(\boldsymbol{x})\boldsymbol{e}_z$. The effective potential

$$\Delta\Omega(h, a, \theta_p) = \Omega(h, a, \theta_p) - \Omega(h = 0, a, \theta_p), \tag{3.61}$$

of the interaction of the particle with the interface is given by the equilibrium value of the functional in Eq. (3.60) for a fixed h:

$$\Omega(h, a, \theta_p) := \min_{\{\mathcal{V}_l\}} \Omega^{sph}_{coex,loc}[h, \{\mathcal{V}_l\}; a, \theta_p]. \tag{3.62}$$

We consider the functional in Eq. (3.62) for the geometry considered in Sects. 3.1-3.6, i.e., in the case when the interface is axially symmetric around the particle and pinned at a cylindrical container at a distance L from the particle. Due to axial symmetry the problem reduces to one dimension and then the functional can be minimized numerically by applying a finite element method (Brakke 1992, see also Chapter 8). We have checked that for the cylinder of radius $L = 5a$ the contributions from the surface integrals at the walls of the container, which also enter the functional, can be safely neglected. In the case $\theta_p = \pi/2$ the dimensionless quantity $\Delta\Omega(h, a, \theta_p = \pi/2)/(\gamma\sigma^2)$ can be directly compared with the macroscopic expression $\tilde{F}(h)$ obtained in Sec. 3.2. The results presented in Fig. 3.9 demonstrate that even for colloidal particles of molecular sizes the effects of long-range intermolecular force have practically no influence on the effective interaction potential of the particle with a liquid-gas interface. Minor asymmetry of the potential with respect to h emerges due to the fact that the interface is actually always locally tangent to the particle surface (there must be a smooth transition to a gas "film" of zero thickness at the particle). For the same reason there

3.6. INFLUENCE OF LONG-RANGE INTERMOLECULAR FORCES

is a non-vanishing deformation of the interface towards the fluid phase even for $h = 0$ (see Fig. 3.9). We note that the model neglects short-range repulsions between the substrate and the fluid molecules. The simplest way to incorporate them would be to introduce a depletion zone of thickness $d_0 \approx \sigma$ along the particle surface, being inaccessible to the fluid particles. More realistic models (i.e., more advanced versions of DFT) could provide a full density profile $\rho(\boldsymbol{r})$ which would then probably yield characteristic oscillations of the fluid density close to the particle surface (compare, e.g., Roth et al. 1999). Furthermore, the fluctuations of the interface could also become important for nanometer-sized particles, nevertheless their influence on the free energy should not be overrated (see, e.g., Cheung & Bon 2009a).

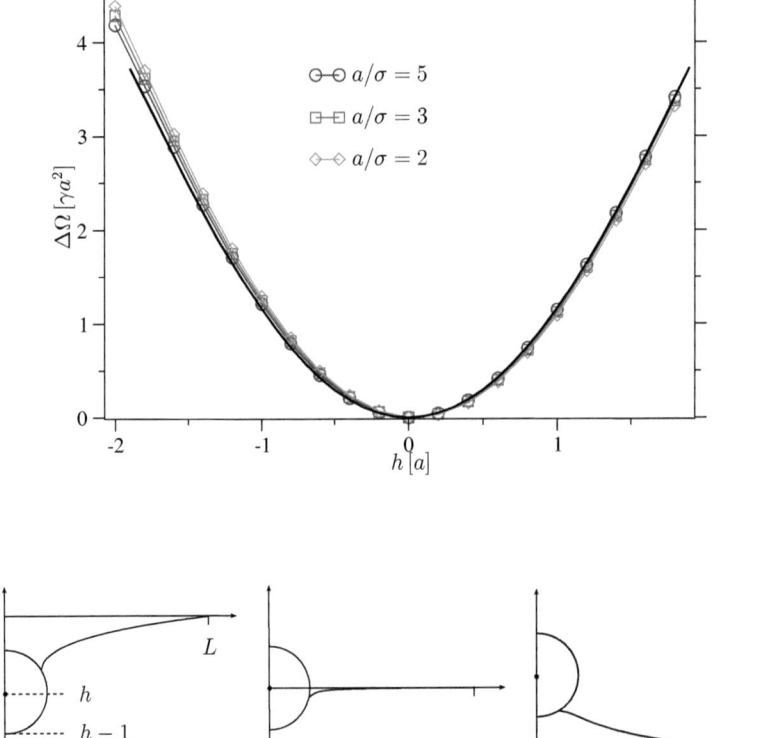

Figure 3.9: The effective potential $\Delta\Omega$ (Eqs. (3.60)-(3.62)) for a spherical nanoparticle of various radii a/σ and with $\theta_p = \pi/2$, at height h above the interface pinned at the distance $L/a = 5$ from the particle. The black solid line represents a part of the corresponding analytic solution $\tilde{F}(h)$ of the macroscopic theory (Eqs. (3.11)-(3.13)) without the metastable branches. The metastable branches of the free energy have not been addressed numerically. The three interface profiles from left to right correspond to $h/a = -1.8, 0$ and 1.8, respectively.

Chapter 4

Capillary interactions

In this Chapter we study the capillary interactions in the generic case of two heavy spherical particles placed at an initially flat interface, under the action of gravity and with the boundary conditions of pinned contact lines at the particles. We derive a functional of the interface profile, which at the minimum yields the free energy $F^{(2)}$. The interaction between the particles is ruled by that part of the free energy which depends on their spatial separation d. The obtained asymptotic result $\Delta F(d) = f^2 \ln(qd)/(2\pi\gamma)$, valid for $qd \ll 1$, is identical as in the case of particles with free contact lines (Oettel et al. 2005b) and does not depend on the particle size a, which indicates that the wetting energy, or in general the mechanism of attachment of the particles to the interface, is irrelevant at large separations. The model with pinned contact lines has the advantage that it enables analytical estimate of the subleading term with the result $O(a/d)^4$ (for general reasons of mechanical equilibrium one can expect a subleading term to be of the the same order also for the case of free contact lines, see the discussion at the end of Sec. 5.2). Particularly, we take into account the tilt ψ (see Fig. 4.1) of the pinned contact lines, which together with their elevation h parameterizes the boundary conditions for the shape of the interface at the particles. Taking only h as a variable is sufficient for calculating the leading term in the free energy, but overestimates the subleading term. The new variable ψ leads to an additional contribution to the free energy (Eq. (4.19)), which then yields the subleading term to be of the order $O(a/d)^4$, in agreement with the general conditions of mechanical equilibrium of the interface (see, c.f., Chapter 5). This complements a recent approach to the effective interactions between capillary multipoles of arbitrary order (Danov et al. 2005), in which the authors assume permanent capillary multipoles, for example dipoles, but do not explain what is their physical origin. The analysis in Chapter 5, which follows the reasoning of Domínguez (2010), shows that the induced capillary monopoles and dipoles must actually vanish if the external forces and torques vanish, respectively.

Furthermore, our approach enables an extension of the superposition approximation by additional fields such that the resulting solution fulfills the boundary conditions at the particles with higher accuracy (see Subsec. 4.2.1). This method resembles the method of reflections (Sec. 4.3) known from Stokes hydrodynamics but in the present context, to our knowledge, it has not yet been considered in the literature.

4.1 Free energy of a heavy particle pinned at the interface

Consider a solid spherical particle of radius a and density ρ_p floating at the liquid-gas interface characterized by the surface tension γ and the difference in densities $\Delta\rho$ between liquid and gas phases. We assume that the three-phase contact line is pinned at the perimeter of the particle, such that only the bottom half of the particle is immersed in the liquid. Physically, the pinning might be due to, for example, roughness of particle surface or different chemical properties of the upper and lower hemispheres (so-called Janus particles (Cheung & Bon 2009b)). This assumption means that the contact line is a circle of radius a, which is rigidly attached to the particle and can move only together with the particle without changing shape. Actually, in such a case the particle can be replaced by a disk, or a vertical cylinder. (The problem of vertical cylinders has been approached by Kralchevsky et al. (1993) but the authors assumed that the cylinders are fixed in their vertical orientations without a possibility to tilt). Accordingly we neglect the surface energy $-\gamma(\cos\theta_p)S_{pl}$ (see Eq. (3.1)) at the particle which is constant in this case. The particle is heavy in the sense that its weight is not balanced by its buoyancy in the reference configuration, which causes the particle to sink into the liquid until the force balance is restored by the emerging vertical capillary force. The free energy functional depending on the interface profile $u(\boldsymbol{x})$, calculated with respect to the reference configuration, can be written as

$$\mathcal{F}[\{u(\boldsymbol{x})\}] = \gamma \int_{\mathbb{R}^2 \setminus A} d^2\boldsymbol{x} \left[\sqrt{1 + (\nabla_\| u)^2} - 1\right] + \Delta\rho g \int_{\mathbb{R}^2 \setminus A} d^2\boldsymbol{x} \int_0^u du' \, u' + \Delta F_{grav} - fh, \quad (4.1)$$

where A is the projection of the region occupied by the particle onto the XY-plane creating a circular disk of radius a. The first term is the surface free energy of the liquid-gas interface and the second term represents the gravitational energy of the displaced liquid under the liquid-gas interface. The third term is a correction due to the gravitational energy of the displaced liquid beneath the particle. The last term is the gravitational energy of the floating particle, where $f = -m^*g$ denotes the effective gravitational force (the minus sign results from taking upward as the positive direction) with $m^* = (4/3)\pi a^3[\rho_p - \Delta\rho/2]$ being the effective mass, according to Archimedes' principle, and h denotes the vertical position of the particle. In the following, we assume that the particle size a is much smaller than the capillary length $\lambda_c = q^{-1} = \sqrt{\gamma/(\Delta\rho g)}$, such that

$$qa \ll 1, \quad (4.2)$$

and therefore we can neglect ΔF_{grav} (this will be justified quantitatively *a posteriori*, see c.f. Eq. (4.15)). Assuming small deformations of the interface, i.e. $|\nabla_\| u| \ll 1$, the Taylor expansion of the free energy functional up to second order in $\nabla_\| u$ reads

$$\mathcal{F}[\{u(\boldsymbol{x})\}] = \gamma \int_{\mathbb{R}^2 \setminus A} d^2\boldsymbol{x} \left[\frac{1}{2}(\nabla_\| u)^2 + \frac{1}{2}q^2 u^2\right] - fh. \quad (4.3)$$

4.1. FREE ENERGY OF A HEAVY PARTICLE

The boundary condition of the contact line pinned at the particle can be expressed as

$$u|_{\partial A} = h, \qquad (4.4)$$

The equilibrium profile is obtained from the condition $\delta \mathcal{F} = 0$ upon variations $\delta u(\boldsymbol{x})$ and δh fulfilling the boundary condition in Eq. (4.4). The total variation $\delta \mathcal{F}$ can be written in the form

$$\delta \mathcal{F} = \gamma \int_{\mathbb{R}^2 \setminus A} d^2\boldsymbol{x} \left[-\nabla_\|^2 u + q^2 u \right] \delta u - \gamma \oint_{\partial A} dl_\| \, (\boldsymbol{n} \cdot \nabla_\| u) \delta u - f \delta h, \qquad (4.5)$$

where $\boldsymbol{n}$ is the outward normal to the boundary ∂A of A in the XY-plane and we have used the divergence theorem together with the fact that the boundary integral vanishes at infinity (owing to the exponential decay of capillary deformations at distances larger than λ_c, see Eq. (2.24)). Vanishing of the first term on the rhs of Eq. (4.5) yields the Euler-Lagrange equation

$$-\nabla_\|^2 u + q^2 u = 0. \qquad (4.6)$$

The boundary condition in Eq. (4.4) implies the relation

$$\delta u|_{\partial A} = \delta h. \qquad (4.7)$$

Thus, the remaining two terms in Eq. (4.5) cancel provided that

$$f = -\gamma \oint_{\partial A} dl_\| \, \boldsymbol{n} \cdot \nabla_\| u, \qquad (4.8)$$

which expresses the force balance on the particle in the vertical direction. According to the general solutions in Eqs. (2.22) and (2.23), the only axially symmetric solution of Eq. (4.6) vanishing at $r \to \infty$ reads

$$\bar{u}(r) = A_0 K_0(qr) \xrightarrow{qr \to 0} -A_0 \ln(qr), \qquad (4.9)$$

where we have used the asymptotic expression for $K_0(x)$ in Eq. (2.25) neglecting the constant $\ln 2 - \gamma_e$. The integration constant A_0 is determined by the boundary condition at $r = a$ (Eq. (4.4)) which finally yields

$$\bar{u}(r) = h \frac{\ln(qr)}{\ln(qa)}. \qquad (4.10)$$

Rewriting Eq. (4.8) by using Eq. (4.8) we have

$$f = -2\pi \gamma a \dot{\bar{u}}(a), \qquad (4.11)$$

where $\dot{\bar{u}}(a) := d\bar{u}/dr|_{r=a}$, and which enables to express the equilibrium solution in terms of system parameters as

$$\bar{u}(r) = -\frac{f}{2\pi \gamma} \ln(qr), \quad qr \ll 1. \qquad (4.12)$$

The governing equation (4.6) together with the divergence theorem can be subsequently applied to transform the surface integral in the free energy functional in Eq. (4.3) into a line integral. The resulting equilibrium free energy can be written as

$$F = -\frac{1}{2}\gamma \oint_{\partial A} dl_\| \left(\boldsymbol{n} \cdot \nabla_\| u\right) u - fh = -\frac{1}{2} fh. \tag{4.13}$$

Inserting $h = \bar{u}(a)$ with the solution (4.12) into Eq. (4.13) yields the expression for the change in the free energy with respect to the reference configuration, which can be interpreted as the *self-energy* of the particle

$$F = -\frac{1}{2} f\bar{u}(a) = \frac{f^2}{4\pi\gamma} \ln(qa) \equiv F_{self}. \tag{4.14}$$

The notion of self-energy will be justified in Chapter 5 by an analogy to the self-energy of a point-charge in electrostatics. However, we note that the expression in Eq. (4.14) does not diverge for $a \to 0$, on contrary to the self-energy of a point-charge. The reason for this is that $|f| = m^*g$ is bounded by the maximal capillary force of the order of γa, independently of the effective mass m^*. If, for example, by increasing the particle density ρ_p the mass m^* would exceed certain maximal value then the particle would detach from the interface. Therefore, as long as the particle is trapped a the interface, the free energy F_{self} vanishes for $a \to 0$ as $a^2 \ln(qa)$. Finally, we estimate ΔF_{grav} in the exact free energy in Eq. (4.1) as

$$\Delta F_{grav} = \Delta\rho g \pi a^2 \int_0^h dh'\, h' = \frac{f^2}{8\pi^2\gamma} \bigl[qa \ln(qa)\bigr]^2. \tag{4.15}$$

By calculating the ratio $\Delta F_{grav}/F_{self} = (1/2)(qa)^2 \ln(qa)$ one sees that ΔF_{grav} can be indeed neglected for $qa \ll 1$.

4.2 Free energy for the case of two particles: effective interactions

In this Section we study the interaction energy of two identical particles placed at $\boldsymbol{x}_1$ and $\boldsymbol{x}_2$, and separated by the distance $d = |\boldsymbol{x}_1 - \boldsymbol{x}_2|$. For simplicity we assume $\boldsymbol{x}_1 = (0,0)$ and $\boldsymbol{x}_2 = (d,0)$. The exact free energy functional for this problem can be written as

$$\mathcal{F}^{(2)}[\{u(\boldsymbol{x})\}] = \gamma \int_{\mathbb{R}^2 \setminus (A_1 \cup A_2)} d^2\boldsymbol{x} \left[\sqrt{1 + (\nabla_\| u)^2} + \frac{1}{2} q^2 u^2\right]$$
$$+ \Delta F_{grav}^{(2)} - f_1 h_1 - f_2 h_2 - F_{ref}^{(2)}, \tag{4.16}$$

where the upper index "(2)" indicates two-particle case. In the following, for simplicity, we treat the particles as identical, i.e., of the same radii and masses, so that $f_1 = f_2 = f$ and as a consequence $h_1 = h_2 = h$ (here, h denotes a variable and not the equilibrium value which in general will be different then in the one-particle problem). Because of

4.2. EFFECTIVE INTERACTIONS

the broken rotational symmetry around each particle, the contact lines can be tilted by an angle ψ (see Fig. 4.1) relative to the planar configuration and therefore we assume that their projections A_i onto XY-plane are in general ellipses and not circular disks. The area of each projection equals the area of an ellipse of the principal axes equal a and $a\cos\psi$, such that

$$|A_1| = |A_2| = \pi a^2 \cos\psi \tag{4.17}$$

Because $A_i \neq A_{i,ref}$, where $A_{1,ref}$ and $A_{2,ref}$ are both circular disks of radius a, the reference free energy

$$F^{(2)}_{ref} = \gamma \int_{\mathbb{R}^2 \setminus (A_{1,ref} \cup A_{2,ref})} d^2x \tag{4.18}$$

cannot be written as an integral over the actual meniscus domain $\mathbb{R}^2 \setminus (A_1 \cup A_2)$ and therefore it has been explicitly separated in Eq. (4.16). The reference configuration corresponds to $g = 0$ and then the free energy does not depend on the positions of the particles, as can be seen from Eq. (4.18). For small deformations of the interface the tilt angle ψ must also be small. Keeping the terms up to $O(\psi^2)$ and $O(u^2)$ and neglecting $\Delta F^{(2)}_{grav} \approx 2\Delta F_{grav}$ (see the discussion after Eq. (4.15)), Eq. (4.16) attains the form

$$\mathcal{F}^{(2)}[\{u(\boldsymbol{x})\}] = \gamma \int_{\mathbb{R}^2 \setminus (A_{1,ref} \cup A_{2,ref})} d^2x \left[\frac{1}{2}(\nabla_\| u)^2 + \frac{1}{2} q^2 u^2\right] + \pi\gamma a^2 \psi^2 - 2fh, \tag{4.19}$$

Due to the tilt of the contact lines, the boundary conditions at the particles differ from Eq. (4.4) and can be expressed as $u|_{\partial A_i} = h + a\sin\psi\cos\phi_i$ which in first order in ψ reads

$$u|_{\partial A_{i,ref}} = h + a\psi\cos\phi_i =: u_{c,i}, \quad i = 1, 2, \tag{4.20}$$

where the angle ϕ_i parameterizes the contact line at particle i. Thus, the variation of the shape of the contact lines is related to the variations of the parameters h and ψ by

$$\delta u_{c,i} = \delta h + a\cos\phi_i \delta\psi. \tag{4.21}$$

The full variation of the free energy functional can be written as

$$\delta\mathcal{F}^{(2)} = \gamma \int_{\mathbb{R}^2 \setminus (A_{1,ref} \cup A_{2,ref})} d^2x \left[-\nabla^2_\| u + q^2 u\right] \delta u$$

$$- 2\left[f + \gamma \oint_{\partial A_{2,ref}} dl_\| (\boldsymbol{n}_2 \cdot \nabla_\| u)\right] \delta h$$

$$+ 2\pi\gamma a^2 \left[-\frac{1}{\pi a} \oint_{\partial A_{2,ref}} dl_\| (\boldsymbol{n}_2 \cdot \nabla_\| u) \cos\phi + \psi\right] \delta\psi$$

$$\tag{4.22}$$

The first term vanishes provided that u obeys the Euler-Lagrange equation in the form in Eq. (4.6). Then the condition $\delta\mathcal{F}^{(2)} = 0$ is satisfied provided that

$$f = -\gamma a \int_0^{2\pi} d\phi_2 \frac{\partial u}{\partial r} \equiv -2\pi\gamma a \left\langle \frac{\partial u}{\partial r} \right\rangle_0 \tag{4.23}$$

$$\psi = \frac{1}{\pi} \int_0^{2\pi} d\phi_2 \frac{\partial u}{\partial r} \cos\phi_2 \equiv 2\left\langle \frac{\partial u}{\partial r} \right\rangle_1, \tag{4.24}$$

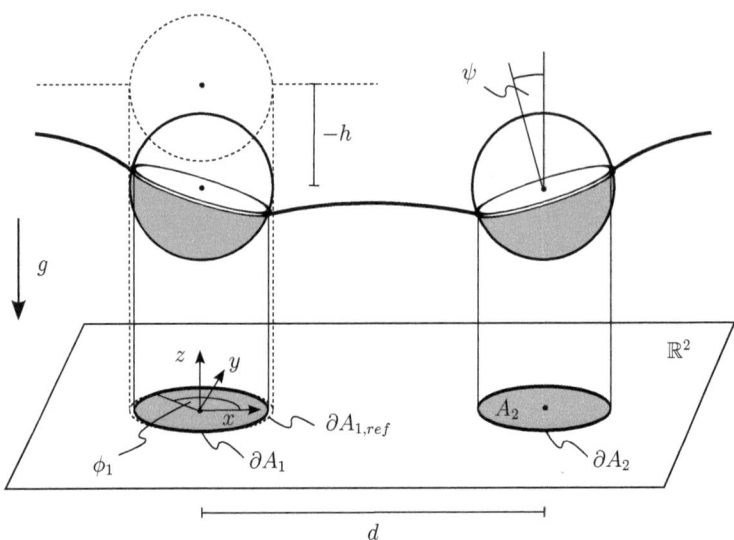

Figure 4.1: Configuration of two heavy particles at the interface and projection of the contact lines onto the XY-plane. Because the contact lines are pinned, the particles be also represented as flat disks. The dashed lines correspond to the reference configuration.

where we introduced the notation $<(\cdot)>_n$ for the n-th moment in the x-direction defined as

$$<(\cdot)>_n =: \frac{1}{2\pi}\int_0^{2\pi} d\phi_2 \cos^n \phi_2(\cdot), \quad n = 0, 1. \tag{4.25}$$

Integrating both sides of Eq. (4.20) over the contact line we also obtain

$$h = <u>_0, \tag{4.26}$$

Using Eq. (4.6) and the divergence theorem the first term in Eq. (4.19) can be transformed into a line integral, and we arrive at the free energy expressed exclusively in terms of the equilibrium deformation u:

$$F^{(2)} = \min_{\{u(\boldsymbol{x})\}} \mathcal{F}^{(2)} = -f<u>_0 + 2\pi\gamma a^2 \left[-\left\langle \frac{(u - <u>_0)}{a}\frac{\partial u}{\partial r}\right\rangle_0 + 2\left(\left\langle \frac{\partial u}{\partial r}\right\rangle_1\right)^2 \right], \tag{4.27}$$

where u obeys Eq. (4.6) with the boundary condition in Eq. (4.20).

4.2.1 Superposition approximation

An exact solution of the capillary equation (4.6) with non-axisymmetric boundary conditions poses a difficult problem, but one can show that an approximate solution

4.2. EFFECTIVE INTERACTIONS

in the form of a superposition of one-particle solutions yields the correct result in the leading order in a/d, i.e., for large separations of the particles. We assume

$$u(\boldsymbol{x}) = u_1(\boldsymbol{x}) + u_2(\boldsymbol{x}), \qquad (4.28)$$

where $u_i(\boldsymbol{x}) \equiv \bar{u}(\boldsymbol{x} - \boldsymbol{x}_i)$ for $i = 1, 2$ are the equilibrium one-particle solutions given by Eq. (4.12). In this approximation the elevation of the contact line at each particle (Eq. (4.26)) equals

$$h = \bar{u}(a) + \bar{u}(d) + \dots. \qquad (4.29)$$

where "..." indicate corrections of higher order in $1/d$. The interaction between the particles is defined by the excess free energy

$$\Delta F = F^{(2)} - 2F_{self}. \qquad (4.30)$$

From the one-particle result $\bar{u}(d) \sim \ln(qd)$, it follows that the derivative $\dot{\bar{u}}(d)$ vanishes like $1/d$ and moreover that each subsequent derivative introduces a factor $1/d$. Thus, the leading order contribution to Eq. (4.27) for large separations, i.e., $d \gg a$, comes only from the first term. At large separations $d \to \infty$ the interaction energy reduces to

$$\Delta F = -f\bar{u}(d) = \frac{f^2}{2\pi\gamma}\ln(qd). \qquad (4.31)$$

Due to the fixed shape of the contact lines, the boundary conditions at the particles cannot be fulfilled exactly by the solution in the form of a simple superposition $u_1 + u_2$. In order to see this, consider the Taylor expansion:

$$(u_1 + u_2)|_{\partial A_2, ref} = \bar{u}(a) + \bar{u}(d) + a\boldsymbol{n}_2 \cdot \nabla_\| \bar{u}|_{de_x} + \frac{1}{2}a^2 \boldsymbol{n}_2 \boldsymbol{n}_2 : \nabla_\| \nabla_\| \bar{u}|_{de_x} + \dots, \qquad (4.32)$$

The gradient $\nabla_\| \bar{u}|_{de_x} = \boldsymbol{e}_x \dot{\bar{u}}(d)$ can be directly related to the tilt angle ψ. Inserting Eq. (4.32) into Eq. (4.24) we obtain $\psi = \dot{\bar{u}}(d)$ and then, using Eq. (4.29), the shape of the contact line at the particle 2 (see Eq. (4.20)) can be written as

$$u_{c,2} = \bar{u}(a) + \bar{u}(d) + a\boldsymbol{n}_2 \cdot \nabla_\| \bar{u}|_{de_x}, \qquad (4.33)$$

such that the Taylor expansion in Eq. (4.32) reads

$$(u_1 + u_2)|_{\partial A_2, ref} = u_{c,2} + \frac{1}{2}a^2 \boldsymbol{n}_2 \boldsymbol{n}_2 : \nabla_\| \nabla_\| \bar{u}|_{de_x} + \dots = u_{c,2} + O(a/d)^2, \qquad (4.34)$$

Thus, the boundary conditions are violated by the terms of the order $O(a/d)^2$. Accordingly, one obtains the following contributions to the excess free energy (Eqs. (4.30), (4.27) and (4.14)):

(i)

$$-f<u>_0 -2F_{self} = -f<u_1+u_2>_0 +f<u_2>_0 = -f<u_1>_0$$
$$= -f\left\langle \bar{u}(d) + a\boldsymbol{n}_2 \cdot \nabla_\| \bar{u}|_{de_x} + \frac{a^2}{2}\boldsymbol{n}_2 \boldsymbol{n}_2 : \nabla_\| \nabla_\| \bar{u}|_{de_x} \right.$$
$$\left. + \frac{a^3}{6}\boldsymbol{n}_2 \boldsymbol{n}_2 \boldsymbol{n}_2 \vdots \nabla_\| \nabla_\| \nabla_\| \bar{u}|_{de_x} + O(a/d)^4 \right\rangle_0 = -f\bar{u}(d) + O(a/d)^4 \qquad (4.35)$$

where we denoted by : and $\vdots$ the second and third rank tensors contractions, respectively, and we used the identity

$$\int_0^{2\pi} d\phi_2 \, \boldsymbol{n}_2 = 0 \qquad (4.36)$$

and

$$\int_0^{2\pi} d\phi_2 \, \boldsymbol{n}_2 \boldsymbol{n}_2 \boldsymbol{n}_2 = 0, \qquad (4.37)$$

as well as

$$\int_0^{2\pi} d\phi_2 \, \boldsymbol{n}_2 \boldsymbol{n}_2 : \nabla_\| \nabla_\| \bar{u}|_{de_x} \propto \int_0^{2\pi} d\phi_2 \, \cos(2\phi) = 0. \qquad (4.38)$$

Similarly, one obtains

(ii)

$$-2\pi\gamma a^2 \left\langle \frac{(u - \langle u \rangle_0)}{a} \frac{\partial u}{\partial r} \right\rangle_0 = -2\pi\gamma a^2 \left\langle \frac{(u_1 - \langle u_1 \rangle_0)}{a} \frac{\partial (u_1 + u_2)}{\partial r} \right\rangle_0$$

$$= -2\pi\gamma a^2 \left\langle \left[\boldsymbol{n}_2 \cdot \nabla_\| \bar{u}|_{de_x} + \frac{a}{2}\boldsymbol{n}_2 \boldsymbol{n}_2 : \nabla_\| \nabla_\| \bar{u}|_{de_x} + \frac{a^2}{6}\boldsymbol{n}_2 \boldsymbol{n}_2 \boldsymbol{n}_2 \vdots \nabla_\| \nabla_\| \nabla_\| \bar{u}|_{de_x} + O(a/d)^4 \right] \times \right.$$

$$\left. \times \left[\dot{\bar{u}}(a) + \boldsymbol{n}_2 \cdot \nabla_\| \bar{u}|_{de_x} + a\boldsymbol{n}_2 \boldsymbol{n}_2 : \nabla_\| \nabla_\| \bar{u}|_{de_x} + O(a/d)^3 \right] \right\rangle_0$$

$$= -\gamma a^2 \dot{\bar{u}}(d)^2 \int_0^{2\pi} d\phi_2 \cos^2 \phi_2 + O(a/d)^4 = -\pi\gamma a^2 \dot{\bar{u}}(d)^2 + O(a/d)^4, \qquad (4.39)$$

where we used Eqs. (4.35)-(4.38); the last contribution is

(iii)

$$4\pi\gamma a^2 \left(\left\langle \frac{\partial u_1}{\partial r} \right\rangle_1 \right)^2 = 4\pi\gamma a^2 \left(\frac{\dot{\bar{u}}(d)}{2\pi} \int_0^{2\pi} d\phi_2 \, \boldsymbol{n}_2 \cdot \boldsymbol{e}_x \cos \phi_2 + O(a/d)^3 \right)^2$$

$$= 4\pi\gamma a^2 \left(\frac{\dot{\bar{u}}(d)}{2\pi} \int_0^{2\pi} d\phi_2 \cos^2 \phi_2 + O(a/d)^3 \right)^2 = \pi\gamma a^2 \dot{\bar{u}}(d)^2 + O(a/d)^4. \qquad (4.40)$$

As can be seen, the sum (i) + (ii) + (iii) yields the contribution to the free energy of the order $O(a/d)^4$. We note that the term (iii), needed to cancel the contribution $-\pi\gamma a^2 \dot{\bar{u}}(d)^2 = O(a/d)^2$ from the term (ii), arises only if one takes into account the tilt of the contact line. The exact form of the correction $O(a/d)^4$ cannot be calculated in the superposition approximation because of the aforementioned errors at the boundaries. To eliminate those errors one has to introduce additional fields, which we shortly discuss in the following Section.

4.3 Method of reflections

The aim of this Section is to indicate that the superposition approximation can be improved to yield a more accurate result. In order to diminish the error at the boundaries we write down the deformation field u in the form

$$u = u_1 + u_2 + u_{12} + u_{21}, \qquad (4.41)$$

where the additional fields u_{12} and u_{21} obey the capillary equation and vanish at infinity. Moreover, we assume that they fulfill the following boundary conditions:

$$u_{12} = -(u_1 + u_2 - u_{c,2}) \quad \text{at} \quad \partial A_{2,ref}, \qquad (4.42)$$
$$u_{21} = -(u_1 + u_2 - u_{c,1}) \quad \text{at} \quad \partial A_{1,ref}. \qquad (4.43)$$

which are of the order of the quadratic term in the expansion in Eq. (4.32). As a consequence, the contributions to the total deformation in Eq. (4.41) from u_{12} propagated back to particle 1 and from u_{21} propagated to particle 2 will be then of the order $O(a/d)^4$. In order to see this, we use Eq. (4.32) to rewrite Eq. (4.42) as

$$u_{12}(r, \phi_2)\Big|_{r=a} = -\frac{1}{2} a^2 \boldsymbol{n}_2 \boldsymbol{n}_2 : \nabla_\| \nabla_\| \bar{u}|_{de_x} = -\frac{f}{4\pi\gamma a} \left(\frac{a}{d}\right)^2 \cos(2\phi_2), \qquad (4.44)$$

from which one can infer that u_{12} (u_{21}) is an "outer" solution (see Subsec. 2.1.5) centered at the particle 2 (1) of order $m = 2$. That means that the "reflected" fields $u_{12}(r)$ and $u_{21}(r)$ are of a quadrupolar type, i.e., they vanish as r^{-2} (with an amplitude $O(a/d)^2$). As a consequence they give a contribution of the order $O(a/d)^4$. Next, the boundary conditions are modified due to those contributions via Eqs. (4.24) and (4.26). Accordingly, h is shifted by $u_{21}|_{de_x}$ and ψ by $\partial_r u_{21}|_{de_x}$. As a consequence, the error at the boundaries is only of the order of $\nabla_\| \nabla_\| u_{21}|_{de_x} = O(a/d)^6$:

$$(u_1 + u_2 + u_{12} + u_{21})|_{\partial A_{2,ref}} = u_{c,2} + O(a/d)^6. \qquad (4.45)$$

with

$$u_{c,2} = \bar{u}(a) + \bar{u}(d) + u_{21}|_{de_x} + a\boldsymbol{n}_2 \cdot \nabla_\|(\bar{u} + u_{21})|_{de_x}. \qquad (4.46)$$

The solution in the form in Eq. (4.41) enables calculation of the contribution $O(a/d)^4$ to the free energy (we do not perform this calculation here). Furthermore, the whole procedure may be iterated by introducing fields u_{121} and u_{212} given by the boundary conditions with the fields u_{21} and u_{12} playing the role of $u_1 + u_2 - u_{c,2}$ and $u_1 + u_2 - u_{c,1}$, respectively, which would enable estimate of the next order corrections (i.e., higher than $O(a/d)^4$) to the free energy. This method is known in Stokesian hydrodynamics as the *method of reflections* (Kim & Karilla 1991), owing the name to the subsequent reflected velocity fields, instead of the two-dimensional deformation field, propagating between the particles. In hydrodynamics, the method is used to calculate the many-body mobility matrix, which represents the linear relation between the velocities of the particles and the given external forces acting on them. In the case of capillary interactions the linearity of the governing equation implies similar linear

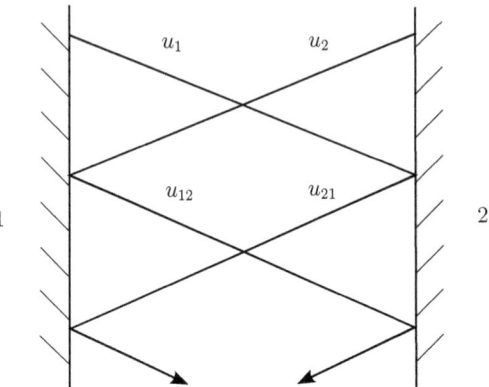

Figure 4.2: Schematic representation of the subsequent reflected fields. The walls represent the surfaces of particles 1 and 2.

relations between the vertical displacements of the particles and given external forces. The advantage of the method of reflections for fixed forces is that the reflected fields are force- and torque-free which means that they vanish as $O(a/d)^4$ and thus give a correction of the order $O(a/d)^4$ to the free energy. In the case of fixed displacements (corresponding, in hydrodynamics, to fixed velocities) reflected fields would not be force-free and would yield a correction of the order $O((\ln(qd))^2/\ln(qa))$. In such a case the accuracy of the asymptotic expression would be much lower. This justifies our choice of the model system in Chapter 7 with the particle at a sessile droplet subjected to a fixed force.

Chapter 5

Electrostatic analogy

In this Chapter we address the analogy between capillarity and electrostatics, developed recently by Domínguez et al. (Domínguez et al. 2008a; Domínguez 2010). Our approach to capillary interactions, based on the concept of an effective surface pressure field and its multipole expansion (see c.f. Chapters 6 and 7), greatly benefits from this analogy and below we present a short derivation of the main results.

5.1 The Poisson equation

As we have shown in Chapter 2, the condition of mechanical equilibrium of an arbitrary piece of the interface can be derived from the Young-Laplace equation. Here, instead, we use the former as a starting point and extend the analysis by allowing an arbitrary external surface pressure field $\Pi(\boldsymbol{r}) + \lambda$, defined on S, which replaces the constant λ in Eq. (2.18). Meanwhile, we assume a vanishing mean-curvature of the interface which corresponds to $\lambda = 0$. In such a case the condition of mechanical equilibrium reads:

$$-\gamma \oint_{\partial S} dl \, \boldsymbol{e}_t = \int_S d^2 S \, \Pi(\boldsymbol{r}) \boldsymbol{e}_n, \qquad (5.1)$$

We neglect gravity in the sense that $\lambda_c \to \infty$, and thus Eq. (5.1) does not contain the term corresponding to the weight of the fluid, but λ_c can still enter the analysis as a large distance cut-off (see, c.f., Eq. (5.13)). In the limit of small deformations of an initially flat interface the versors $\boldsymbol{e}_n$ and $\boldsymbol{e}_t$ (Eqs. (2.13) and (2.12) with $z(r) \equiv u(r)$) read

$$d^2 S \boldsymbol{e}_n = dx \, dy [-\nabla_\parallel u + \boldsymbol{e}_z] \qquad (5.2)$$

and

$$dl\boldsymbol{e}_t = dl_\parallel \boldsymbol{n} \cdot \left[\mathbf{1} + \boldsymbol{e}_z \nabla_\parallel u - \nabla_\parallel u \nabla_\parallel u + \frac{1}{2} (\nabla_\parallel u)^2 \mathbf{1} \right] + O(\nabla_\parallel u)^4. \qquad (5.3)$$

Accordingly, the z-component of Eq. (5.1) up to first order in $\nabla_\parallel u$ attains the form

$$-\gamma \oint_{\partial A} dl_\parallel \boldsymbol{n} \cdot \nabla_\parallel u = \int_A dA \, \Pi(\boldsymbol{x}). \qquad (5.4)$$

Applying the divergence theorem to the line integral in the above equation we obtain the linearized Young-Laplace equation in the form

$$-\nabla_\parallel^2 u = \frac{1}{\gamma}\Pi, \tag{5.5}$$

which is an analogue of the Poisson equation in $2D$-electrostatics with u playing the role of the electrostatic potential, Π playing the role of the charge distribution ("capillary charge") and γ being electric permittivity. The analogy is complete up to the reversal of the sign of the capillary force. Two-dimensional electric charge distribution $\rho(\boldsymbol{x})$ in an external potential $\phi(\boldsymbol{x})$, where $\boldsymbol{x} \in A$ would experience the force $-\int_A dA\,\rho\nabla_\parallel\phi$. The lateral capillary force acting on the capillary charge distribution $\Pi(\boldsymbol{x})$ is given by minus the lateral component of Eq. (5.1), which reads $(1/\gamma)\int_A dA\,\Pi\nabla_\parallel u$. Thus, it has the reverse sign of what the electrostatic analogy would imply. The reason for this is that the attraction of like capillary charges (while the opposite is true for electric charges) is in fact due to an external agent Π, and not only due to capillary forces itself.

The usual boundary conditions on u also have a close analogy to electrostatics:

(i) imposing a fixed deformation of the contact line $u_0(\boldsymbol{x})$ at a boundary ∂A, such that

$$u(\boldsymbol{x})\big|_{\partial A} = u_0(\boldsymbol{x}), \tag{5.6}$$

corresponds to the Dirichlet boundary condition (given potential);

(ii) a prescribed contact angle $\theta_0(\boldsymbol{x})$, for example with a vertical wall,

$$\boldsymbol{n}\cdot\nabla_\parallel u(\boldsymbol{x})\big|_{\partial A} = \cot\theta_0(\boldsymbol{x}), \tag{5.7}$$

corresponds to the Neumann boundary condition. For consistency with the assumption of small deformations of the interface the angle $\theta_0(\boldsymbol{x})$ should be everywhere close to $\pi/2$.

We note that the linear theory works in the regions of low curvature even if there are other regions where non-linearities become important (Domínguez et al. 2008a). The deformation at the boundaries of those non-linear "patches" provide boundary conditions for the regions where the linear theory is valid. Inversely, one can always find a virtual "capillary charge" distribution inside the "patches" corresponding to the given boundary conditions (see, c.f., Eq. (5.19)). Thus, in analogy to Gauß' theorem in electrostatics, the total capillary charge Q_0 inside a region A, can be defined as (see Eq. (5.4))

$$Q_0 := -\gamma\oint_{\partial A} dl_\parallel\,\boldsymbol{n}\cdot\nabla_\parallel u, \tag{5.8}$$

On the other hand the vertical component of the capillary force according to Eq. (5.4) counterbalances the total external vertical force $f = \int_A dA\,\Pi(\boldsymbol{x})$ acting at the interface inside the region A and we obtain

$$Q_0 = f. \tag{5.9}$$

In the same manner it can be shown (Domínguez et al. 2008a) that the total capillary dipole Q_1 (see, c.f., Eq. (5.17)) is related to the total torque m, exerted by the external force, by

$$Q_1 = e_z \times m. \tag{5.10}$$

5.2 Green's function and multipole expansion

The general solution for the deformation u governed by Eq. (5.5) can be written as

$$u(\boldsymbol{x}) = \frac{1}{\gamma} \int d^2\boldsymbol{x}' \, \Pi(\boldsymbol{x}') G(\boldsymbol{x}, \boldsymbol{x}') \tag{5.11}$$

where G is the Green's function of the $2D$ Laplace equation

$$-\nabla_\parallel^2 G(\boldsymbol{x}, \boldsymbol{x}') = \delta(\boldsymbol{x} - \boldsymbol{x}'). \tag{5.12}$$

Assuming isotropy and uniformity of the interface one has $G(\boldsymbol{x}, \boldsymbol{x}') = G(|\boldsymbol{x} - \boldsymbol{x}'|)$ and the solution of Eq. (5.12) reads (see, e.g., Nehari 1975)

$$G(r) = -\frac{1}{2\pi} \ln\left(\frac{r}{\zeta}\right). \tag{5.13}$$

where the constant ζ reflects the dependence of the Green's function on specific boundary conditions (for example, in presence of gravity $\zeta = \lambda_c$). In terms of the complex variable $z = x + iy$ the deformation u can be expressed in terms of the complex potential $V(z)$, such that $u(\boldsymbol{x}) = \operatorname{Re} V(z)$, where

$$V(z) := -\frac{1}{2\pi\gamma} \int d^2\boldsymbol{x}' \, \Pi(z') \ln\left(\frac{z-z'}{\zeta}\right) \tag{5.14}$$

Assume that the pressure field Π is localized inside a region of size a around $z' = 0$, i.e., $\Pi(z') = 0$ for $|z'| > a$. Provided that the point of observation $z = r\exp(i\phi)$ lies outside this region such that $r > a$, the following Taylor expansion can be used (Nehari 1975):

$$\ln(z - z') = \ln z - \sum_{n=0}^{\infty} \frac{1}{n}\left(\frac{z'}{z}\right)^n. \tag{5.15}$$

Inserting this into Eq. (5.14) gives

$$V(z) = \frac{\tilde{Q}_0}{2\pi\gamma} \ln\left(\frac{\zeta}{z}\right) + \frac{1}{2\pi\gamma} \sum_{n=1}^{\infty} \frac{\tilde{Q}_n}{nz^n}, \tag{5.16}$$

where $\tilde{Q}_n$ are the complex capillary multipoles defined as

$$\tilde{Q}_n := \int d^2\boldsymbol{x}' \, \Pi(z') z'^n, \tag{5.17}$$

which can be also written as $\tilde{Q}_n = Q_n e^{i\varphi_n}$, where $Q_n = |\tilde{Q}_n|$ are the real multipoles. The phases $\varphi_n = \arg \tilde{Q}_n$ can be interpreted as orientations of the multipoles. By applying the residue theorem to Eq. (5.16) it can be shown that all the multipoles $\tilde{Q}_n$ are fully determined by the deformation around an arbitrary contour C enclosing the origin (Domínguez 2010):

$$\tilde{Q}_n = i\gamma \oint_C dz\, z^s \frac{dV}{dz}, \qquad (5.18)$$

which after applying Cauchy-Riemann relationships verified by the analytic function $V(z)$ leads to

$$Q_n = \gamma \left| \oint_C dl_{\parallel} (x + iy)^n (\boldsymbol{n} - i\boldsymbol{e}_z \times \boldsymbol{n}) \cdot \nabla_{\parallel} u \right|, \qquad (5.19)$$

where $\boldsymbol{n}$ is a normal to C in the xy-plane pointing outwards the area enclosed by C.

The free energy of an initially flat, unbounded interface exposed to an arbitrary pressure $\Pi(\boldsymbol{x})$ in the limit of small deformations reads

$$\begin{aligned} F = \int d^2x \left[\frac{\gamma}{2} (\nabla_{\parallel} u)^2 - \Pi(\boldsymbol{x}) u(\boldsymbol{x}) \right] &= -\frac{1}{2} \int d^2x\, \Pi(\boldsymbol{x}) u(\boldsymbol{x}) \\ &= -\frac{1}{2\gamma} \int d^2x \int d^2x'\, \Pi(\boldsymbol{x}) G(\boldsymbol{x}, \boldsymbol{x}') \Pi(\boldsymbol{x}'), \end{aligned} \qquad (5.20)$$

where we have used Eq. (5.5) and the boundary conditions of a vanishing deformation at infinity. Thus, we obtained an expression analogical to the potential energy of a charge distribution in electrostatics, but with a minus sign, which reflects the already mentioned property that like capillary charges attract each other. The multipole expansion in Eq. (5.16) can be applied to calculate interaction free energy in Eq. (5.20) of two sources Π_1 and Π_2 separated by the distance d. Assume that the pressure field Π is of the form

$$\Pi(\boldsymbol{x}) = \Pi_1(\boldsymbol{x}) + \Pi_2(\boldsymbol{x} - d\boldsymbol{e}_x), \qquad (5.21)$$

where $\Pi_i(\boldsymbol{x}) = 0$ for $|\boldsymbol{x}| > a$ (for simplicity we assume both pressure distributions to be of equal spacial extent a). Inserting this $\Pi(\boldsymbol{x})$ into the expression in Eq. (5.20) the free energy splits into three parts

$$F = F_{1,self} + F_{2,self} + \Delta F \qquad (5.22)$$

where $F_{i,self}$ represent, in analogy to electrostatics, self-energies

$$F_{i,self} = -\frac{1}{2\gamma} \int d^2x \int d^2x'\, \Pi_i(\boldsymbol{x}) G(\boldsymbol{x}, \boldsymbol{x}') \Pi_i(\boldsymbol{x}') \quad i = 1, 2 \qquad (5.23)$$

and ΔF is the interaction energy given by

$$\begin{aligned} \Delta F &= \frac{1}{2\pi\gamma} \int d^2x \int d^2x'\, \Pi_1(\boldsymbol{x}) G(\boldsymbol{x}, \boldsymbol{x}' + d\boldsymbol{e}_x) \Pi_2(\boldsymbol{x}' \\ &= -\frac{1}{\gamma} \mathrm{Re} \int d^2x \int d^2x'\, \Pi_1(z) \ln\left(\frac{z - z' - d}{\zeta} \right) \Pi_2(z'). \end{aligned} \qquad (5.24)$$

Using the Taylor expansion in Eq. (5.15) and the identity

$$(z - z')^n = \sum_{k=0}^{n} \frac{n!}{k!(n-k)!} z^n z'^{n-k}, \qquad (5.25)$$

we obtain the interaction energy ΔF in the form of the multipole expansion

$$\Delta F = -\frac{1}{\gamma} \sum_{n=0}^{\infty} \sum_{n'=0}^{\infty} Q_{1,n} Q_{2,n'} G_{nn'}(d) \cos(n\varphi_{1n} + n'\varphi_{2n'}), \quad (5.26)$$

where the matrix elements $G_{nn'}(d)$ read

$$G_{nn'}(d) = \frac{1}{2\pi} \frac{(-1)^n |n + n' - 1|!}{n! n'!} \times \begin{cases} \ln\left(\frac{\zeta}{d}\right) & \text{for } n = n' = 0, \\ \frac{1}{d^{n+n'}} & \text{otherwise} \end{cases} \quad (5.27)$$

Thus, the interaction between multipoles n, n' behaves as $\ln(\zeta/d)$ for $n = n' = 0$ and vanishes as $(a/d)^{n+n'}$ for $n \neq 0$ or $n' \neq 0$. Moreover, according to the expression in Eq. (5.27) we can estimate the interaction free energy of heavy colloidal particles addressed in Chapter 4. Due to $Q_0 = f \neq 0$ the particles correspond to capillary monopoles so that the leading term in the free energy is $-Q_0^2 \ln(\lambda_c/d)/(2\pi\gamma)$ with $Q_0 = f$, which recovers the result in Eq. (4.31). The capillary dipoles associated with the particles vanish due to vanishing external torques ($Q_1 = 0$) and therefore the next non-vanishing moment is the quadrupole Q_2, which yields the subleading term in the interaction energy $\sim Q_0 Q_2 / d^2$. For a single particle, due to axial symmetry, we have $Q_n = 0$ for $n \geq 2$, so that in the case of two particles those multipoles can only be induced by the deformation field due to the second particle. Particularly, Q_2 is induced by a quadrupolar component of this field (i.e., by u_{12} or u_{21}, see Sec. 4.3), which means that $Q_2 = Q_0 \times O(a/d)^2$ (however, this fact cannot be inferred directly from the electrostatic analogy) and we obtain a correction to the free energy of the order $Q_0^2 \times O(a/d)^4$.

5.3 Mechanical equilibrium of the interface

In any experimental setup the interface is always bounded by walls of a container. At distances larger than the capillary length λ_c the deformation of the interface around a particle vanishes exponentially whereas at distances smaller than λ_c it exhibits a logarithmic behavior (Eq. (4.9)). In the language of electrostatics capillary deformation is screened due to gravity for $d > \lambda_c$. This has an important consequence for the balance of forces acting at the interface. Any force applied to the interface at $r = 0$ is counterbalanced by a force due to the hydrostatic pressure $-\Delta \rho g u = -\gamma u / \lambda_c^2$ integrated over the interface for distances $r < \lambda_c$ (see Eq. (2.27)). However, for $r \ll \lambda_c$ the contribution from the hydrostatic pressure can be neglected and the mechanical equilibrium demands that all the external forces acting on the boundaries of the interface must cancel. Thus, for example, the force exerted at the interface by the walls of the container must balance the force acting on the particle associated with the capillary charge Q_0. Mathematically this fact can be expressed as

$$Q_{\text{wall}} = -\gamma \oint_{\text{wall}} dl_\parallel \, \boldsymbol{n} \cdot \nabla_\parallel u = -Q_0, \quad (5.28)$$

where Q_{wall} is the capillary charge of the container. As a conclusion, in the case of a system size smaller than the capillary length the presence of boundaries cannot be neglected (for the case of a particle at a droplet see Domínguez et al. 2007a). Moreover, in the absence of gravity (for example in experiments in weightlessness, see Langbein 2002), the capillary length becomes infinite and then the shape of the liquid interface is determined exclusively by the shape of the walls and by the contact angle.

5.4 Method of images

As an example of the effect of boundaries we study the behavior of a particle characterized by a capillary charge Q_0 positioned at $x = d$ next to a vertical wall at $x = 0$. We assume that in the absence of the particle the interface is flat and the contact angle at the wall is $\theta_0 = \pi/2$. The shape of the interface in the presence of the particle will be determined by the kind of boundary conditions at the wall, which typically are (i) free contact line and fixed contact angle $\theta(\boldsymbol{x}) = \theta_0$ or (ii) pinned contact line. As already mentioned those two cases correspond to (i) Neumann and (ii) Dirichlet boundary conditions in electrostatics. One can exploit the electrostatic analogy in order to solve the boundary value problem by introducing a virtual image charge Q' at the virtual continuation of the reference interface beyond the wall (see Fig. 5.1). The position and sign of the image at the virtual semi-plane should be chosen such that the superposition of the deformations from the particle and its image satisfies the proper boundary conditions at the wall. For symmetry reason it is easy to guess that in both cases the image should be placed at $x = -d$. The image charge Q' should be chosen as follows:

(i) $Q' = Q_0$, such that the contributions to the slope of the interface $\boldsymbol{n} \cdot \nabla_\| u$ coming from the actual charge and its image cancel each other at the wall and the contact angle remains unchanged. This can be seen by symmetry between the one particle solution $u_1(|\boldsymbol{x} - d\boldsymbol{e}_x|)$ for the original particle and $u_2(|\boldsymbol{x} + d\boldsymbol{e}_x|)$ for its image with respect to the reflection in the plane $x = 0$. Because of the equal capillary charges, we have $u_2(r) = u_1(r) = \bar{u}(r)$. Then, if we write the deformation in the form of the superposition $u = \bar{u}(|\boldsymbol{x} - d\boldsymbol{e}_x|) + \bar{u}(|\boldsymbol{x} + d\boldsymbol{e}_x|)$, then one obtains $\boldsymbol{n} \cdot \nabla_\| u|_{\boldsymbol{x}=(0,y)} = 0$

(ii) $Q' = -Q_0$, such that the contributions to the total deformation at the wall cancel each other and as a result the deformation at the contact line vanishes. Indeed, in this case we have $u_2(r) = -u_1(r) = -\bar{u}(r)$, so that for the solution in the form $u = \bar{u}(|\boldsymbol{x} - d\boldsymbol{e}_x|) - \bar{u}(|\boldsymbol{x} + d\boldsymbol{e}_x|)$ we obtain $u|_{\boldsymbol{x}=(0,y)} = 0$.

The capillary image forces were first studied by Kralchevsky et al. (1994) who has also taken into account the gravitational potential energy of the particle. If the contact angle is slightly different than $\pi/2$ the initial interface is tilted and as a consequence the particle feels an additional effective potential (attractive or repulsive depending on the interplay between contact angle, weight and buoyancy of the particle). As a consequence, in special cases, the total potential composed of the surface free energy

5.4. METHOD OF IMAGES

and the gravitational potential energy can be a non-monotonic function of d. In the case of a heavy particle (weight prevailing buoyancy), $\theta_0 > \pi/2$ and the contact line pinned at the wall, the total potential exhibits a minimum at the distance where gravity is balanced by the capillary force. In the case of the same particle, but $\theta_0 < \pi/2$ and a free contact line, the potential exhibits a maximum. We are going to observe a qualitatively similar non-monotonic behavior of the effective potential as a function of the separation from the contact line for a particle at a sessile droplet with a free or a pinned contact line at the substrate, however, with the non-monotonicity being in this case exclusively due to the curvature of the droplet surface and the volume constraint (and not due to an external potential).

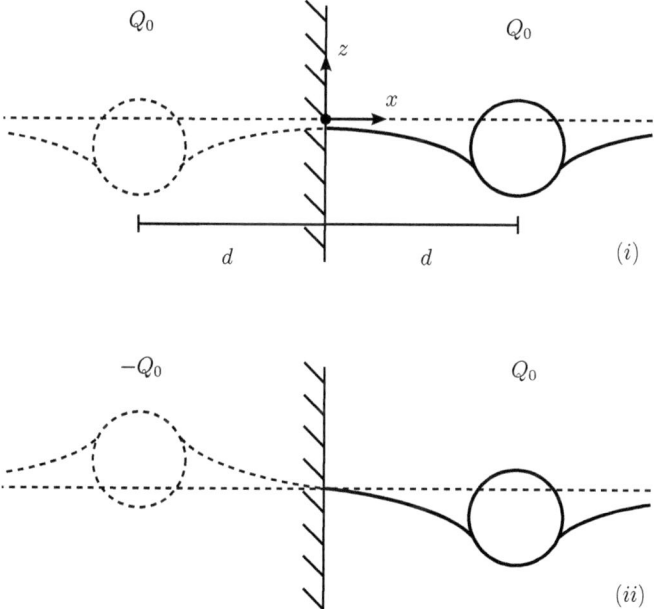

Figure 5.1: Capillary monopole Q_0 (for example due to the effective mass m^* in the case of a heavy particle) next to a wall and its image, depending on the boundary conditions at the wall: (i) a free contact line, (ii) a pinned contact line.

As a last remark we note that the method of images remains valid also in the case when the gravitational energy of the fluid cannot be neglected, i.e. at distances $d > \lambda_c$. The method of images can still be applied, because the governing equation is modified only by adding a term $q^2 u$ (see Eq. (4.6)) to the lhs of Eq. (5.5) and therefore it remains linear. As a consequence, the solution can be constructed as a superposition of single-particle solutions for the original particle and its image. Thus, in the case of a heavy particle with the effective mass m^* at the distance $d \sim \lambda_c$ from a vertical wall, the boundary condition of a fixed contact angle corresponds to the image particle of the same mass m^* at the virtual semi-plane, also at a distance d from the wall. Similarly, the boundary condition of a pinned contact line corresponds to the image particle of mass $-m^*$.

Chapter 6

Deformations of a spherical droplet

Unlike a flat interface, the interface of a spherical liquid droplet constitutes a closed surface. In the case of an incompressible liquid the constant volume of the droplet imposes an additional constraint on the shape of the interface. In presence of an external perturbation the surface tension acts such that it tends to restore the spherical shape of the droplet, which is the one minimizing the surface area for a given volume. This restoring force plays a similar role as gravity in the case of a free flat interface. As a consequence the droplet radius R_0 plays the role of the capillary length in the sense that the pointlike perturbation leads to deformation vanishing as $\ln(R_0/r)$ for distances $r \ll R_0$. This holds as long as R_0 is smaller than the actual capillary length $\lambda_c = \sqrt{\gamma/(\Delta \rho g)}$. Otherwise the effects of gravity have to be taken into account.

Here, we present an analysis of the deformations of a full spherical (non-sessile) droplet for $R_0 < \lambda_c$. Our theory is based on the approach of Morse & Witten (1993), who studied the problem of compressibility of emulsions. Inside an emulsion the deformation of a single droplet is due to contact with the surrounding droplets. The surface pressure field describing the effect of these interactions has always negative sign, because the droplet can only be pushed, but not pulled, by the other droplets. In our analysis we allow both signs, which is necessary when the perturbation is due to a colloidal particle. This is a natural extension of the Morse-Witten theory and does not yield any additional theoretical difficulties. Furthermore, we perform an expansion of the deformation and the pressure fields in terms of spherical harmonics $Y_{lm}(\Omega)$, in which the expansion coefficients are numbered by the degree l and order m. In the case when the pressure field is due to a small colloidal particle, we show that the coefficients with $n := |m| = 0$ are associated with an external force acting on the particle and those with $n = 1$ with an external torque. Thus, in general, a force- and torque-free particle corresponds to an effective pressure field characterized by $n = 2$, in full analogy to the case of a flat interface. Finally, the expressions for the interaction potentials between two sources, characterized by orders n and n', are obtained in closed forms for arbitrary angular separations by summing over all the degrees l. Particularly, by matching the amplitudes of the deformation with the asymptotic case of a flat interface ($R_0 \to \infty$), we obtain an analytical expression for the interaction potential between two nearly spherical ellipsoidal particles trapped at the interface of a droplet.

6.1 Surface free energy of a spherical droplet.

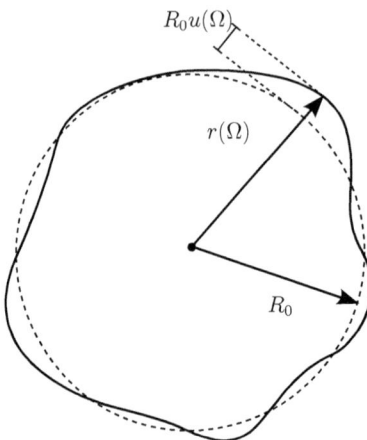

Figure 6.1: Deformations $r(\Omega)$ of a spherical droplet of initial radius R_0. The difference defines small deformations $u(\Omega) = r(\Omega) - R_0$.

Consider a spherical droplet of radius R_0 subjected to an external surface pressure field $\Pi(\Omega)$ parameterized by spherical coordinates $\Omega = (\theta, \phi)$ on the unit sphere. The equilibrium shape of the droplet can be found by minimizing the free energy functional $\mathcal{F}$ expressed in terms of the radial position of the interface $r(\Omega) = R_0 + u(\Omega)$ and consisting of the surface free energy minus the work done by the external pressure Π in displacing the interface (with respect to the reference state $R_{ref}(\Omega) \equiv R_0$). The volume constraint can be incorporated by adding a term $-\lambda(V - V_l)$ with the sign of the Lagrange multiplier $-\lambda$ chosen such that λ (at equilibrium) can be identified with the internal pressure of the droplet, and where the liquid volume reads $V_l = 4\pi R_0^3/3$. The volume $V = V[\{u(\Omega)\}]$ can be expressed as an integral over the whole sphere of the infinitesimal volume element $(1/3)[(R_0 + u)^3 - R_0^3]d\Omega$, and then one arrives at the free energy functional in the form

$$\mathcal{F}[\{r(\Omega)\}] = \int d\Omega \left(\gamma \left[s(u, \nabla_a u) - R_0^2 \right] - \frac{1}{3}\left[\lambda + \Pi(\Omega)\right]\left[(R_0 + u)^3 - R_0^3\right] \right), \quad (6.1)$$

with $d\Omega = d\theta d\phi \sin\theta$ and

$$s(u, \nabla_a u) = (R_0 + u)^2 \sqrt{1 + (\nabla_a u)^2/(R_0 + u)^2}, \quad (6.2)$$

where

$$\nabla_a := \mathbf{e}_\theta \partial_\theta + \frac{\mathbf{e}_\phi}{\sin\theta} \partial_\phi \quad (6.3)$$

6.1. SURFACE FREE ENERGY OF A SPHERICAL DROPLET.

is the dimensionless angular gradient on the unit sphere. The volume constraint $V = V_l$ reads

$$\int d\Omega \left[(R_0 + u)^3 - R_0^3\right] = 0. \tag{6.4}$$

All the above expressions are valid without additional approximations as long as the interface has no overhangs and they provide a starting point to the linear analysis in terms of small deformations $v = u/R_0$. In the following, we assume that the interfacial gradients are small, i.e., $|\nabla_a v| \ll 1$ everywhere on the droplet, such that we also have $v \ll 1$. We introduce dimensionless quantities: the external pressure $\pi = \Pi R_0/\gamma$ and the shift $\mu = \lambda R_0/\gamma - 2$ in the internal pressure relative to the dimensionless Laplace pressure of a perfectly spherical droplet equal to 2. In order to perform the expansion of the free energy functional in v we note that, as soon as a vanishing external pressure leads to a vanishing deformation, π and μ must be $O(v)$. In the zeroth order in v the interface is in the reference configuration and thus the free energy functional $\mathcal{F}$ is identically zero. In first order in v it also vanishes, which can be explained by the fact that the spherical droplet is in the stable equilibrium and thus linear variations of the shape give no contribution to the free energy. Thus, the first non-vanishing contribution comes in second order and one obtains

$$\frac{1}{\gamma R_0^2}\mathcal{F}[\{v(\Omega)\}] = \int d\Omega \left[\frac{1}{2}(\nabla_a v)^2 - v^2 - \left(\pi(\Omega) + \mu\right)v + O(v^3)\right]. \tag{6.5}$$

Applying the stationary condition $\delta \mathcal{F}/\delta v \to 0$ to Eq. (6.5) leads to the Euler-Lagrange equation in the form

$$-\left(\nabla_a^2 + 2\right)v(\Omega) = \pi(\Omega) + \mu. \tag{6.6}$$

Note that in Eq. (6.5) the Lagrange multiplier μ is $O(v)$, such that the volume constraint in Eq. (6.4) must be satisfied also only in first order in v (and not necessarily in higher orders). Thus, one obtains the condition

$$\int d\Omega\, v = 0. \tag{6.7}$$

Using Eq. (6.6) we can calculate the free energy as

$$\frac{1}{\gamma R_0^2} F = \frac{1}{\gamma R_0^2} \min_{\{v(\Omega)\}} \mathcal{F} = \int d\Omega \left[\frac{1}{2}\nabla_a(v\nabla_a v) - \frac{1}{2}v\left(\nabla_a^2 v + 2v\right) - \left(\pi(\Omega) + \mu\right)v\right]$$

$$= -\frac{1}{2}\int d\Omega\left(\pi(\Omega) + \mu\right)v = -\frac{1}{2}\int d\Omega\, \pi v, \tag{6.8}$$

where in the fourth equality we integrated by parts and used the fact that the interface is closed and the integral $\int d\Omega\, \nabla_a(v\nabla_a v)$ vanishes due to the divergence theorem. The last equality follows from the volume constraint in Eq. (6.7).

6.2 Expansion in spherical harmonics.

We expand both the deformation $v(\Omega)$ and the pressure $\pi(\Omega)$ in spherical harmonics $Y_{lm}(\Omega)$,

$$v(\Omega) = \sum_{lm} v_{lm} Y_{lm}^*(\Omega), \tag{6.9}$$

$$\pi(\Omega) = \sum_{lm} \pi_{lm} Y_{lm}^*(\Omega). \tag{6.10}$$

where v_{lm} and π_{lm} are the expansion coefficients, subsequently called multipoles, and read

$$v_{lm} := \int d\Omega'\, v(\Omega') Y_{lm}(\Omega'), \tag{6.11}$$

$$\pi_{lm} := \int d\Omega'\, \pi(\Omega') Y_{lm}(\Omega'). \tag{6.12}$$

In terms of the multipoles the governing equation (6.6) turns into an infinite set of algebraic equations (Morse & Witten 1993):

$$[l(l+1) - 2] v_{lm} = \pi_{lm} + \mu \delta_{l0}, \tag{6.13}$$

where $l = 0, 1, \ldots$ and $m = -l, \ldots, l$. The volume constraint in Eq. (6.7) implies that the $l = 0$ component of v vanishes, i.e., $v_{00} = 0$ and as a consequence

$$\mu = -\pi_{00}, \tag{6.14}$$

which means that the internal pressure shift counterbalances the external pressure. It also follows from Eq. (6.13) that the $l = 1$ components of the deformation v are undefined. The reason for this is that those components describe translations of the whole droplet without any change in shape which do not change the free energy. On the other hand $l = 1$ components of the external pressure π must all cancel, which reflects the condition of balance of forces acting on the droplet. Indeed, according to Eq. (6.12), the multipoles $\pi_{1-1}, \pi_{10}, \pi_{11}$ are related to the Cartesian components f_x, f_y, f_z of the total force $\boldsymbol{f}$ acting on the droplet by

$$\pi_{10} \equiv \sqrt{\frac{3}{4\pi}} \int d\Omega'\, \pi(\Omega') \cos\theta' = \sqrt{\frac{3}{4\pi}} f_z, \tag{6.15}$$

$$\pi_{1\pm 1} \equiv \sqrt{\frac{3}{4\pi}} \int d\Omega'\, \pi(\Omega') \sin\theta' e^{\pm i\phi'} = \sqrt{\frac{3}{4\pi}} (f_x \pm i f_y). \tag{6.16}$$

Hence, the condition $\boldsymbol{f} = 0$ is equivalent to

$$\pi_{1m} = 0, \quad m = -1, 0, 1. \tag{6.17}$$

Obviously, a free droplet subjected to a pointlike external force (corresponding to a capillary monopole) is not in mechanical equilibrium. This fact must be taken into account in deriving Green's function describing the deformation of the interface in response to a pointlike force.

In mechanical equilibrium the external pressure π acting at an infinitesimal piece of a spherical interface can only exert a force in the radial direction. Therefore there is no torque with respect to the droplet center. However, one can consider any point at the interface as a reference, and then, with respect to this point, there is a non-vanishing torque $\boldsymbol{m}$ associated with π. Without loosing generality we can choose this point to lie at the z-axis, and then the torque $\boldsymbol{m}$ is related to multipoles π_{1-1} and π_{11} by

$$\pi_{1\pm1} = \sqrt{\frac{3}{4\pi}} \int d\Omega'\, \pi(\Omega') \sin\theta' e^{\pm i\phi'} \equiv \sqrt{\frac{3}{4\pi}}(-m_y \pm im_x). \qquad (6.18)$$

This shows that, in fact, the condition of balance of forces acting on the whole droplet as expressed in Eq. (6.17) implies the torque-balance with respect to an arbitrary point at the interface, which might be, for example the center of a colloidal particle.

6.3 Green's function

In the case of N pointlike forces acting on the interface only those configurations are allowed in which the total force vanishes. Consider a pressure field in the form of superposition of the delta functions centered at N different directions Ω_i

$$\pi(\Omega) = \sum_{i=1}^{N} q_i \delta(\Omega, \Omega_i), \qquad (6.19)$$

where $q_i := f_i/(\gamma R_0)$ are the dimensionless forces and $\delta(\Omega, \Omega_i) = \delta(\theta-\theta_i)\delta(\phi-\phi_i)/\sin\theta_i$ is the Dirac delta distribution expressed in terms of spherical coordinates. The total pressure field must obey the condition of balance of forces which in analogy to Eq. (6.17) might be expressed as

$$\sum_{i}^{N} \pi_{i,1m} = 0, \quad m = -1, 0, 1. \qquad (6.20)$$

The corresponding interface deformation can be written as a superposition of single particle contributions:

$$v(\Omega) = \sum_{i}^{N} q_i G(\Omega, \Omega_i), \qquad (6.21)$$

where $G(\Omega, \Omega')$ is the Green's function describing a response of the interface at a direction Ω to a point-force applied at a direction Ω' and according to Eqs. (6.6), (6.14) and (6.19)-(6.21) it obeys the equation

$$-(\nabla_a^2 + 2)G(\Omega, \Omega') = \sum_{l \geq 2} \sum_{m=-l}^{l} Y_{lm}^*(\Omega) Y_{lm}(\Omega'), \qquad (6.22)$$

where the right hand side is a modified Dirac delta function $\hat{\delta}(\Omega - \Omega')$ with the $l = 0$ and $l = 1$ components projected out. Integrating both sides of Eq. (6.26) over the unit sphere one sees that the $l = 0$ component of G vanishes, which reflects the condition of constant volume

$$\int d\Omega\, G(\Omega, \Omega') = 0. \tag{6.23}$$

The $l = 1$ component of G also vanishes, which can be seen by multiplying both sides of Eq. (6.26) by a radial vector $\boldsymbol{e}_r$, integrating over the unit sphere and using the fact that $\boldsymbol{e}_r$ can be decomposed into spherical harmonics with $l = 1$ exclusively. This reflects the assumption that the center of mass of the droplet is fixed in space, which can be written as $\int d\Omega\, \boldsymbol{e}_r G(\Omega, \Omega') = 0$ or

$$\int d\Omega\, Y_{1m}(\Omega) G(\Omega, \Omega') = 0, \quad m = -1, 0, 1, \tag{6.24}$$

where the z-axis is chosen arbitrarily. G can be also decomposed into spherical harmonics and then Eq. (6.22) yields the following expression:

$$G(\Omega, \Omega') = \sum_{l \geq 2} \sum_{m=-l}^{l} \frac{1}{l(l+1) - 2} Y_{lm}^*(\Omega) Y_{lm}(\Omega') = \frac{1}{4\pi} \sum_{l \geq 2} \frac{2l+1}{l(l+1) - 2} P_l(\cos \bar{\theta}), \tag{6.25}$$

where we have used the summation rule for spherical harmonics and $\bar{\theta}$ denotes the angle between the directions Ω and Ω'. By using identities for infinite series containing Legendre polynomials (Prudnikov et al. 1986a) one obtains G in a closed form (Morse & Witten 1993):

$$G(\bar{\theta}) = -\frac{1}{4\pi} \left[\frac{1}{2} + \frac{4}{3} \cos \bar{\theta} + \cos \bar{\theta} \ln \left(\frac{1 - \cos \bar{\theta}}{2} \right) \right]. \tag{6.26}$$

We have already pointed out that the pressure field in the form of a single point-force violates balance of forces expressed in Eq. (6.20). Therefore, if we assume that the solution is in the form of the Green's function, then it must be associated with an additional pressure field counterbalancing the point-force. Inserting

$$v(\Omega) = G(\Omega, \Omega') \tag{6.27}$$

into Eq. (6.6) and using Eq. (6.22), we obtain the corresponding pressure field π in the form

$$\pi(\Omega) = \delta(\Omega, \Omega') + \pi_{CM}(\Omega, \Omega'), \tag{6.28}$$

with the effective pressure

$$\pi_{CM}(\Omega, \Omega') = -\frac{3}{4\pi} \cos \bar{\theta}. \tag{6.29}$$

which must emerge in order to cancel $l = 1$ component of the external pressure $\pi_{ext}(\Omega) = \delta(\Omega, \Omega')$. Therefore, π_{CM} corresponds to a body force fixing the center of mass (CM) of the droplet. As can be seen from Eqs. (6.27) and (6.28) the contribution from π_{CM} to the deformation field v vanishes by construction. This can also be checked explicitly by using Eq. (6.24):

$$\int d\Omega'\, \pi_{CM}(\Omega') G(\Omega, \Omega') \propto \int d\Omega'\, Y_{1m}(\Omega') G(\Omega, \Omega') = 0. \tag{6.30}$$

6.4. CAPILLARY INTERACTIONS.

Finally, we note that for small angular distances, according to Eq. (6.26) one has $G(\bar{\theta}) \to (1/2\pi)\ln\bar{\theta}$, which yields the deformation due to a point-force in the form

$$v(\bar{\theta}) = -\frac{1}{2\pi}\ln\left(\frac{r}{R_0}\right) \quad (6.31)$$

where $r = \bar{\theta}R_0$ is the arc length and R_0 plays the role of the capillary length. However, one has to remember that the asymptotic form in Eq. (6.31) is valid only for a droplet with a fixed center of mass (this issue has recently raised controversy in the literature, see Würger 2006a; Domínguez *et al.* 2007a) or immobilized by other means (e.g. by being attached to a substrate).

The expression for the deformation under the action of point-forces in Eq. (6.21) can be generalized to the case of an arbitrary continuous force distribution $\pi_{ext}(\Omega)$,

$$v(\Omega) = \int d\Omega'\, \pi_{ext}(\Omega') G(\Omega, \Omega'), \quad (6.32)$$

The total pressure π acting at the interface has an additional component responsible for fixing the center of mass, i.e., $\pi = \pi_{ext} + \pi_{CM}$. Inserting the expression in Eq. (6.32) into Eq. (6.8) we can express the free energy in terms of the Green's function as

$$\frac{1}{\gamma R_0^2}F = -\frac{1}{2}\int d\Omega \int d\Omega'\, \pi_{ext}(\Omega) G(\Omega, \Omega') \pi_{ext}(\Omega'), \quad (6.33)$$

where the contribution from π_{CM} vanishes by the virtue of Eq. (6.30).

6.4 Capillary interactions.

Assume that the pressure π_{ext} is localized around two different directions Ω_1 and Ω_2 and introduce a following decomposition:

$$\pi_{ext}(\Omega) = \pi_1(\hat{R}_1^{-1}\Omega) + \pi_2(\hat{R}_2^{-1}\Omega), \quad (6.34)$$

where $\hat{R}_1$ denotes the rotation transforming the original coordinate frame xyz into the coordinate frame $x'y'z'$, referred to as O_1, associated with the pressure distribution π_1 and $\hat{R}_2$ denotes the rotation transforming the original coordinate frame xyz into the coordinate frame $x''y''z''$, referred to as O_2, associated with the pressure distribution π_2 (see Fig. 6.2). We use the parameterization in terms of Euler angles, in which the rotation $\hat{R}$ is a composition of the rotation around the z-axis by the angle α, around the (rotated) y-axis by the angle β and finally around the (rotated) z-axis by the angle γ. This is equivalent (see, e.g., Hamermesh 1962) to the rotation around fixed (not rotated) axes as follows: around the z-axis by the angle γ, around the y-axis by the angle β and finally around the z-axis by the angle α. Under the rotation $\hat{R}(\alpha,\beta,\gamma)$ of the coordinate frame the coordinates transform under $\hat{R}^{-1}$, which we have symbolically indicated in Eq. (6.34).

The total free energy can be written as $F = F_{1,self} + F_{2,self} + \Delta F$, where $F_{i,self} = -f_i^2/(4\pi\gamma)\ln(R_0/a_i) + O(1)$ is the self-energy of particle i, which does not depend on the relative position of the particles on the droplet. The interaction free energy ΔF is given by the cross-terms in Eq. (6.33) with π given by Eq. (6.34),

$$\frac{1}{\gamma R_0^2}\Delta F = -\int d\Omega \int d\Omega'\, \pi_1(\hat{R}_1^{-1}\Omega) G(\Omega,\Omega') \pi_2(\hat{R}_2^{-1}\Omega')$$
$$= -\int d\Omega \int d\Omega' \sum_{lm_1} \pi_{1,lm_1} Y^*_{lm_1}(\hat{R}_1^{-1}\Omega) \sum_{jm} g_j Y^*_{jm}(\Omega) Y_{jm}(\Omega') \sum_{km_2} \pi_{2,km_2} Y^*_{km_2}(\hat{R}_2^{-1}\Omega'), \tag{6.35}$$

where we have used the decomposition of G as obtained in Eq. (6.25) with

$$g_l = \begin{cases} 0 & \text{for } l = 0, 1 \\ \dfrac{1}{l(l+1)-2} & \text{for } l \geq 2. \end{cases} \tag{6.36}$$

Without loosing generality we can assume that $\Omega_1 = 0$ and that the reference frame O_1 coincides with the global reference frame xyz, which implies that $\hat{R}_1$ is an identity operator. In such a case Eq. (6.35) reduces to

$$\frac{1}{\gamma R_0^2}\Delta F = -\int d\Omega' \sum_{l \geq 2, m_1} \pi_{1,lm_1} Y^*_{lm_1}(\Omega')\, g_l \sum_{k \geq 2, m_2} \pi_{2,km_2} Y^*_{km_2}(\hat{R}_2^{-1}\Omega'), \tag{6.37}$$

where we have used the orthogonality of spherical harmonics, i.e., $\int d\Omega Y_{lm}(\Omega) Y^*_{l'm'}(\Omega) = \delta_{ll'}\delta_{mm'}$. Spherical harmonics transform under the representation of the group of rotations according to Wigner (1959)

$$Y_{lm}(\hat{R}^{-1}\Omega) = \sum_{m=-l}^{l} D^l_{m',m}(\hat{R}) Y_{lm'}(\Omega), \tag{6.38}$$

where $D^l_{m',m}$ is the Wigner D-matrix and reads (here, we use the convention of Brink & Satchler 1968)

$$D^l_{m',m}(\hat{R}) = D^l_{m',m}(\alpha,\beta,\gamma) = e^{-im'\alpha} d^l_{m',m}(\beta) e^{-im\gamma}, \tag{6.39}$$

where $d^l_{m',m}(\beta)$ is known as the Wigner (small) d-matrix (and subsequently simply referred to as the Wigner matrix) and we have used an opposite sign convention. We use the parameterization in terms of the orientations ϕ_1 and ϕ_2 of coordinate frames O_1 and O_2, respectively, with respect to the geodesic Λ_{12} connecting the points $\Omega_1 = 0$ and Ω_2 on the unit sphere (see Fig. 6.2). The rotation $\hat{R}_2$ is then parameterized by the triad of the Euler angles $(2\pi - \phi_1, \theta_{12}, \phi_2)$ and we obtain

$$Y^*_{km_2}(\hat{R}_2^{-1}\Omega') = \left[\sum_{m'=-k}^{k} D^k_{m',m_2}(\hat{R}_2) Y_{km'}(\Omega')\right]^*$$
$$= \sum_{m'=-k}^{k} (-1)^{m_2-m'} D^k_{-m',-m_2}(2\pi - \phi_1, \theta_{12}, \phi_2)(-1)^{m'} Y_{k,-m'}(\Omega')$$
$$= (-1)^{m_2} \sum_{m'=-k}^{k} d^k_{-m',-m_2}(\theta_{12}) e^{i(-m'\phi_1 + m_2\phi_2)} Y_{k,-m'}(\Omega') \tag{6.40}$$

6.4. CAPILLARY INTERACTIONS.

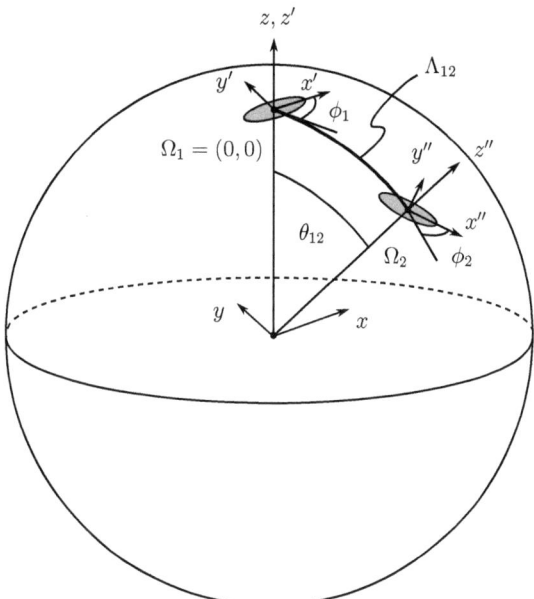

Figure 6.2: Angular configuration of the pressure distributions symbolically represented as gray ellipses centered at the directions $\Omega_1 = (0,0)$ and Ω_2 on a unit sphere. The angles ϕ_1 and ϕ_2 represent orientations of the local coordinate frames associated with the pressure distributions (see main text). The thick solid line denotes the geodesic Λ_{12} from Ω_1 to Ω_2.

where we have used the identities $Y^*_{lm}(\Omega) = (-1)^m Y_{l,-m}(\Omega)$ and $D^l_{m',m}{}^*(\alpha,\beta,\gamma) = (-1)^{m-m'} D^l_{-m',-m}(\alpha,\beta,\gamma)$. Applying the transformation in Eq. (6.38) to Eq. (6.35) one finally obtains:

$$\frac{1}{\gamma R_0^2}\Delta F = -\sum_{l\geq 2}\sum_{m_1=-l}^{l}\sum_{m_2=-l}^{l} \pi_{1,lm_1}\pi_{2,lm_2}(-1)^{m_2} g_l\, d^l_{m_1,-m_2}(\theta_{12})\, e^{i(m_1\phi_1+m_2\phi_2)}. \quad (6.41)$$

6.4.1 Limit of small particles

Having in a perspective colloidal particles as sources of the effective surface pressure π, we would like to relate multipoles π_{lm}, $l \geq 0$, $m = -l, \ldots, l$ to a prescribed force, torque and higher capillary multipoles Q_n, $n \geq 0$, defined for the case of a flat interface. When $a \ll R_0$ the interface can be treated as locally flat in a close neighborhood $\Delta\Omega$ of angular size $O(a/R_0)$ around each particle separately. If $\pi(\Omega)$ vanishes outside $\Delta\Omega$ each spherical multipole π_{lm}, being a complex number, can be expressed in terms of a modulus $Q_{|m|}$ and a phase $\tilde{\phi}_{|m|}$. Approximating $\Delta\Omega$ by a circular disk $D(a)$ of radius a

centered at the origin of the tangent plane and using the asymptotic form of spherical harmonics
$$Y_{lm}(\theta,\phi) \xrightarrow[\theta \to 0]{} i^{|m|+m} A_{l|m|} \theta^{|m|} e^{im\phi}, \qquad (6.42)$$

where
$$A_{ln} = \sqrt{\frac{2l+1}{4\pi} \frac{(l+n)!}{(l-n)!} \frac{1}{2^n n!}}, \quad n > 0, \qquad (6.43)$$

we obtain

$$\pi_{i,lm} = \int_{\Delta\Omega} d\Omega' \, \pi_i(\Omega') Y_{lm}(\Omega')$$

$$\xrightarrow[\omega \to 0]{} i^{|m|+m} A_{l|m|} \times \left\{ \begin{array}{l} \int_{D(a)} \frac{d^2x'}{R_0^2} \frac{R_0 \Pi_i(z')}{\gamma} \left(\frac{z'}{R_0}\right)^{|m|}, \quad m \geq 0 \\ \left[\int_{D(a)} \frac{d^2x'}{R_0^2} \frac{R_0 \Pi_i(z')}{\gamma} \left(\frac{z'}{R_0}\right)^{|m|}\right]^*, \quad m < 0 \end{array} \right\} =$$

$$= i^{|m|+m} A_{l|m|} \left(\frac{a}{R_0}\right)^{|m|+1} Q_{i,|m|} e^{im\tilde{\phi}_{i,|m|}}, \qquad (6.44)$$

where z' denotes a complex number and $\Pi_i(\boldsymbol{x}') \equiv \gamma\pi_i(\Omega'(\boldsymbol{x}'))/R_0$, with $\boldsymbol{x}'(\Omega')$ being the projection onto the plane tangent to the reference sphere at Ω_i. We have also defined the dimensionless quantities

$$Q_{i,n} := \left| \int_{D(a)} \frac{d^2x'}{a^2} \frac{a\Pi_i(z')}{\gamma} \left(\frac{z'}{a}\right)^n \right|, \qquad (6.45)$$

and

$$\tilde{\phi}_{i,n} := \frac{1}{n} \arg \int_{D(a)} \frac{d^2x'}{a^2} \frac{a\Pi_i(z')}{\gamma} \left(\frac{z'}{a}\right)^n. \qquad (6.46)$$

where $n \geq 0$. In Eq. (6.45) $Q_{i,n}$ is nothing but a capillary multipole defined in Eq. (5.17)) for a flat interface. Accordingly, due to the conditions of the force and torque balance on a flat interface (see Domínguez et al. 2008b), $Q_{i,0}$ can be identified as a "capillary monopole", i.e., the total external force acting on the particle, $Q_{i,1}$ as a "capillary dipole", i.e., the total external torque. Accordingly, in the case of a free particle, $Q_{i,2}$ is the lowest non-vanishing multipole. Eq. (6.44) shows that the multipole order $n \equiv |m|$ decides about the physical properties of the source not only in the case of a flat interface but also in the case of a spherical interface. Furthermore, we note that for each particle i one can always choose the orientation ϕ_i of the coordinate frame O_i such that the phase of the order n vanishes, i.e., $\tilde{\phi}_{i,n} = 0$. Finally, we can write the interaction energy $\Delta F_{nn'}$ for a pair of capillary multipoles $Q_{1,n}$ and $Q_{2,n'}$ of arbitrary orders $n \geq 0$ and $n' \geq 0$ as a sum of terms with $m_1 = \pm n$ and $m_2 = \pm n'$ in Eq. (6.41). According to Eq. (6.44) we have $\pi_{l-n} = (-1)^n \pi_{ln}$ and by using the property of the Wigner matrix $d_{m',m}^l = (-1)^{m-m'} d_{m,m'}^l = d_{-m,-m'}^l$ we obtain

6.4. CAPILLARY INTERACTIONS.

$$\frac{1}{\gamma R_0^2}\Delta F_{nn'} \xrightarrow[a_{1,2}/R_0 \to 0]{} -Q_{1,n}Q_{2,n'}\frac{a_1^{n+1}a_2^{n'+1}}{R_0^{n+n'+2}}\sum_{l\geq 2} A_{ln}\,A_{ln'}\,g_l$$

$$\times \begin{cases} 2\left[(-1)^{n'} d_{n,-n'}^l(\theta_{12})\cos(n\phi_1+n'\phi_2) \right. \\ \qquad \left. + d_{n,n'}^l(\theta_{12})\cos(n\phi_1-n'\phi_2)\right], & n>0,\ n'>0, \\ 2\,d_{n,0}^l(\theta_{12})\cos(n\phi_1), & n>0,\ n'=0, \\ d_{0,0}^l(\theta_{12}), & n=0,\ n'=0. \end{cases}$$

(6.47)

Thus, for $a_1 = a_2 = a$ the interaction free energy scales with the droplet radius R_0 as

$$\Delta F_{nn'} \sim \gamma R_0^2 \left(\frac{a}{R_0}\right)^{n+n'+2} = \gamma a^2 \left(\frac{a}{R_0}\right)^{n+n'}. \qquad (6.48)$$

Note, that in the case of two monopoles one has $n + n' = 0$ and then the interaction free energy does not depend on R_0 but only on $Q_{1,0}$ and $Q_{2,0}$.

6.4.2 Explicit formulas for interaction potentials.

In this Section we evaluate the interaction energy in Eq. (6.47) for pairs of multipoles of the same order $n = n'$.

Monopoles, $n = n' = 0$.

From Eq. (6.47) together with Eqs. (6.25) and (6.43), and by using the identity $d_{0,0}^l(\theta_{12}) = P_l(\cos\theta_{12})$ we obtain

$$\Delta F_{00}(\theta_{12}) = -\gamma a^2 Q_{1,0} Q_{2,0} \sum_{l\geq 2} \frac{2l+1}{4\pi[l(l+1)-2]} P_l(\cos\theta_{12}) \equiv -\gamma a^2 Q_{1,0} Q_{2,0} G(\theta_{12}).$$

(6.49)

which is consistent with the general expression in Eq. (6.33) for the pressure field in the form $\pi_{ext} = Q_{1,0}\delta(\Omega,\Omega_1) + Q_{2,0}\delta(\Omega,\Omega_2)$. The condition of the mechanical equilibrium of the droplet implies that, unless Ω_1 and Ω_2 point in the exactly opposite directions and $Q_{1,0} = Q_{2,0}$, the center of mass of the droplet must be fixed by an external body force.

Quadrupoles, $n = n' = 2$.

In this case we obtain

$$\Delta F_{22}(\theta_{12},\phi_1,\phi_2) = -2\gamma a^2 \left(\frac{a}{R_0}\right)^4 Q_{1,2}Q_{2,2}$$
$$\times \sum_{l\geq 2}\left(A_{l2}\right)^2 g_l \left[\cos(2\phi_1+2\phi_2)\,d_{-2,2}^l(\theta_{12}) + \cos(2\phi_1-2\phi_2)\,d_{2,2}^l(\theta_{12})\right], \quad (6.50)$$

where ϕ_1 and ϕ_2 are the orientations of the quadrupoles measured as indicated in Fig. 6.2 and chosen such that the complex phases $\tilde{\phi}_{i,n}$ for $n = 2$ vanish. In Appendix A we show that the multiplicative coefficient in front of $\cos(2\phi_1 - 2\phi_2)$ vanishes and that the interaction energy simplifies to

$$\Delta F_{22}(\theta_{12}, \phi_1, \phi_2) = -\frac{3\gamma a^2}{64\pi}\left(\frac{a}{R_0}\right)^4 Q_{1,2}Q_{2,2}\cos(2\phi_1 + 2\phi_2)\sin^{-4}\frac{\theta_{12}}{2}. \quad (6.51)$$

In the limit $\theta_{12} \ll 1$ the interface between the particles can be approximated as flat and then the arc length $R_0\theta_{12}$ can be identified with their direct spatial separation d. Thus, one can compare our result with the known result for two *ellipsoidal particles* at a flat interface (Stamou et al. 2000; Fournier & Galatola 2002; Lehle et al. 2008), which in the case of two identical particles reads (Lehle et al. 2008)

$$\Delta F_{el,flat} = -3\pi\gamma(\Delta u_{max})^2 \left(\frac{a}{d}\right)^4 \cos(2\phi_1 + 2\phi_2), \quad (6.52)$$

where the undulation of the contact line at the particle $\Delta u_{\max}$ is the maximal difference in the vertical displacement of the points at the contact line. Eq. (6.52) is valid under the assumption that the shape of the particles only slightly deviates from spherical. Under such an assumption, matching the solutions in Eqs. (6.51) and (6.52) in the limit $R_0 \to \infty$ with $a/R_0\theta_{12} \simeq a/d = const \ll 1$ yields the quadrupole moment Q_2 of an ellipsoidal, almost-spherical particle in the form

$$Q_2 = 2\pi\Delta u_{max}/a, \quad (6.53)$$

As a consequence, the interaction energy between two identical ellipsoidal particles at a spherical interface can be written as

$$\Delta F_{22}(\theta_{12}, \phi_1, \phi_2) = -\frac{3\pi}{16}\gamma(\Delta u_{max})^2 \left(\frac{a}{R_0}\right)^4 \cos(2\phi_1 + 2\phi_2)\sin^{-4}\frac{\theta_{12}}{2}. \quad (6.54)$$

Applicability of the above formula is limited to large enough angular separations θ_{12} such that Δu_{max} can be treated as a constant independent of θ_{12}.

Chapter 7

Particles at the surface of sessile droplets

In this Chapter we study the free energy of a spherical particle at the surface of a *sessile* droplet. We investigate the influence of the substrate and derive the corresponding boundary conditions for small deformations of the droplet surface from a free energy functional containing the substrate surface free energy. The actual equilibrium free energy turns out to be independent of the position of the particle on the droplet as long as the particle is force-free and then the shape of the droplet is a perfect cap of sphere (referred to as the reference configuration). However in the case when an external force is applied to the particle in the radial direction (with respect to the origin determined by the center of the reference cap of sphere), the free energy depends on the angular position of the particle on the droplet. In such a case the condition of balance of forces acting on the droplet in the lateral direction requires either a fixed lateral position of the center of mass of the droplet (model A) or a pinned contact line at the substrate (model B).

Both situations can occur in experiments. For example, in the presence of gravity a small tilt of the substrate could provide a lateral body-force counterbalancing the lateral component of $f\mathbf{e}_r$. (However, in this case also the deformation of the droplet due to a non-zero normal component of the gravitational force must be taken into account.) On the other hand, pinning of the contact line can occur due to heterogeneities of the substrate. To some extent such a pinning is always present in actual systems because the substrate is never perfectly smooth. Therefore, in practical terms, we expect this latter case to be the more relevant one. These two different conditions for mechanical equilibrium impose distinct additional constraints onto the equilibrium shape of the droplet and thus onto the minimization of the free energy functional in Eq. (7.3). In the limit of small deformations the forthcoming detailed analysis will reveal that those two cases correspond to Robin and Dirichlet boundary conditions for the deformation field at the substrate, respectively.

In the non-axisymmetric cases we are limited to the perturbation theory, within which we derive an approximate free energy functional (see Eq. (7.23) and Appendix B). The effects of the particle pulled by an external force f and of fixed center of mass are incorporated by introducing the effective pressure fields $\pi(\Omega)$ and $\pi_{CM}(\Omega)$, which enter the linear Young-Laplace equation governing the small deformations of the droplet (Eq. (7.26)). We show that in the limit of small particles $a/R_0 \to 0$ the free energy of the sessile droplet (Eq. (7.51)), expressed in terms of Green's function satisfying the boundary conditions at the substrate corresponding to either a free or a pinned contact line (Eqs. (7.48) and (7.49)), do not depend on the size of the particle but only on the pulling force f, the contact angle θ_0 at the substrate, and on the angular position of the particle α. In the special case $\theta_0 = \pi/2$ we obtain analytic expressions for the Green's functions by using the method of images (Eqs. (7.64) and (7.67)). Finally, we calculate the pair-potentials for two particles and show that the resulting equilibrium configurations depend sensitively on the boundary conditions at the substrate (Figs. 7.7 and 7.8).

7.1 Single particle

We consider a smooth spherical particle of radius a trapped at the surface of a liquid droplet of volume V_l residing on a planar substrate. The equilibrium contact angles θ_0 at the substrate and θ_p at the particle are given by Young's law:

$$\cos\theta_0 = \frac{\gamma_{0g} - \gamma_{0l}}{\gamma}, \qquad (7.1)$$

$$\cos\theta_p = \frac{\gamma_{pg} - \gamma_{pl}}{\gamma}, \qquad (7.2)$$

where γ_{ab} are surface tensions with the indices $a, b = 0, p, l, g$ standing for substrate, particle, liquid, and gas, respectively, and $\gamma_{lg} \equiv \gamma$. In polar coordinates, the position of the particle can be described by the radial displacement h and the polar angle α (see Fig. 7.1).

It can be shown (Kralchevsky & Nagayama 2001) that in the absence of external forces the particle adjusts its immersion, independently of its lateral position at the droplet, according to Young's law, such that the droplet remains an undeformed spherical cap. The configuration with the particle at the unperturbed droplet will be called the reference configuration (see Fig. 7.1), for which we set $h = 0$. As a consequence of the spherical shape of the droplet the corresponding free energy does not depend on the polar angle α.

7.1.1 Free energy functional

If, however, an external force f acts on the particle in radial direction, the interface deforms such that the capillary force counterbalances the external force. The corresponding equilibrium shape of the droplet minimizes the following free energy functional:

$$\begin{aligned}\mathcal{F}[\{\boldsymbol{r}(\Omega)\}, h, \alpha; f, \{\gamma_{ab}\}, a, V_l, \lambda] = \\ = \gamma(S_{lg} - S_{lg,ref}) - \gamma\cos\theta_0(S_{0l} - S_{0l,ref}) - \gamma\cos\theta_p(S_{pl} - S_{pl,ref}) \\ - fh - \lambda(V - V_l), \quad (7.3)\end{aligned}$$

7.1. SINGLE PARTICLE

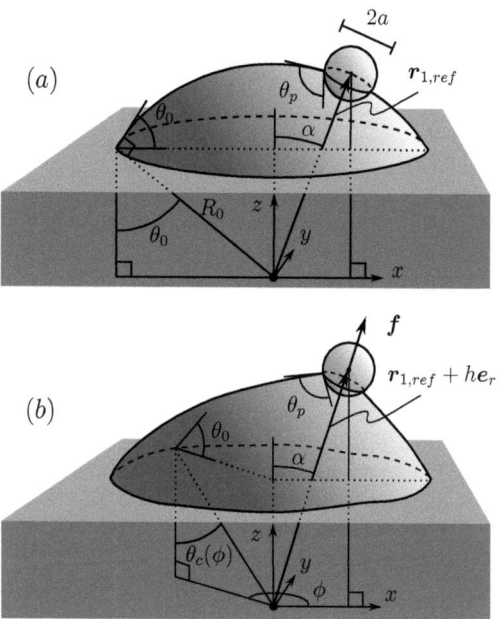

Figure 7.1: Sketch of the system (see main text); (a) reference configuration; (b) deformation due to an external force $\boldsymbol{f} = f\boldsymbol{e}_r$ for a free contact line at the substrate. In this case the contact angle at the substrate equals θ_0 along the contact line. The direction of the x-axis is chosen such that the center of the particle at $\boldsymbol{r}_{1,ref} + h\boldsymbol{e}_r$ (with $h=0$ in (a)) lies in the xz plane. In the case of a pinned contact line at the substrate (not shown), for $f \neq 0$ and $\alpha \neq 0$ the contact angle at the substrate varies along the contact line (i.e., it is a function of ϕ). The contact line at the particle is free, which means a constant contact angle equal to θ_p.

where $\{\boldsymbol{r}(\Omega)\}$ denotes the configuration of the interface in terms of spherical coordinates $\Omega = (\theta, \phi)$ on the unit sphere, and S_{lg}, S_{0l}, S_{pl} are the liquid-gas, substrate-liquid, and particle-liquid surface areas, respectively. In Eq. (7.3) $\mathcal{F}$ is the free energy functional relative to the reference configuration of the unperturbed droplet referred to by the index ref. The last term with the Lagrange multiplier λ ensures that the liquid volume is constant. The sign of this term has been chosen such that the value of λ minimizing the functional in Eq. (7.3) can be interpreted as the internal pressure. The sizes of the droplet (R_0) and of the particle (a) are considered to be sufficiently large so that in Eq. (7.3) we can neglect line tension contributions and Eqs. (7.1) and (7.2) do indeed hold (Schimmele et al. 2007).

The condition of mechanical equilibrium of the droplet in the presence of an external force requires a mechanism fixing the overall lateral position of the droplet on the substrate. The substrate provides a counterbalancing force in the vertical direction, but not in the lateral direction, and therefore the droplet would start to move under the action of the external force. Thus, mechanical equilibrium is only possible in the case that a body-force fixes the center of mass of the droplet or, alternatively, that the contact line is pinned at the substrate. Both situations can occur in experiments. For example, in the presence of gravity a small tilt of the substrate could provide a lateral body-force counterbalancing the lateral component of $f\mathbf{e}_r$. (However, in this case also the deformation of the droplet due to a non-zero normal component of the gravitational force must be taken into account.) On the other hand, pinning of the contact line can occur due to heterogeneities of the substrate. To some extent such a pinning is always present in actual systems because the substrate is never perfectly smooth. Therefore, in practical terms, we expect this latter case to be the more relevant one. These two different conditions for mechanical equilibrium impose distinct additional constraints onto the equilibrium shape of the droplet and thus onto the minimization of the free energy functional in Eq. (7.3). We shall use an index σ in order to distinguish between the cases of a free ($\sigma = A$) or a pinned ($\sigma = B$) contact line, respectively.

A fixed lateral position x_{CM} of the center of mass (CM) can be achieved by adding a term $-f_{CM}(x_{CM} - x_{CM,ref})$ to the free energy functional $\mathcal{F}$, where $x_{CM,ref}$ is determined by the reference configuration and $-f_{CM}$ is a Lagrange multiplier. Choosing the minus sign in the prefactor $-f_{CM}$ will enable us to identify f_{CM} with the force acting at the center of mass and fixing it. We introduce the free energy $F_A^\star$ as a constrained minimum of $\mathcal{F}$, with α kept constant and with a fixed center of mass:

$$F_A^\star(\alpha; f, \theta_0, \theta_p, a, R_0, \lambda, f_{CM}) = \min_{\{r(\Omega)\}, h} \left[\mathcal{F} - f_{CM}(x_{CM} - x_{CM,ref}) \right], \quad (7.4)$$

where $R_0 = R_0(\theta_0, \theta_p, a, V_l)$ is the radius of the droplet in the reference configuration. In the following, if not indicated otherwise, we shall suppress the explicit dependence on the set $\{f, \theta_0, \theta_p, a, R_0\}$ of independent system parameters. The Lagrange multipliers λ and f_{CM} can be determined as functions of these independent system parameters from the conditions

$$V(\alpha; \lambda, f_{CM}) = V_l, \quad (7.5)$$

$$x_{CM}(\alpha; \lambda, f_{CM}) = x_{CM,ref}. \quad (7.6)$$

This renders $\lambda = \lambda(\alpha; f, \theta_0, \theta_p, a, R_0)$ and $f_{CM} = f_{CM}(\alpha; f, \theta_0, \theta_p, a, R_0)$ which upon insertion into $F_A^\star(\alpha; f, \theta_0, \theta_p, a, R_0, \lambda, f_{CM})$ yields $F_A(\alpha; f, \theta_0, \theta_p, a, R_0)$.

On the other hand, the condition of a pinned contact line imposes only a geometric constraint onto the minimization of the free energy, which we symbolically indicate as

$$F_B^\star(\alpha) = \min_{\{r(\Omega)\}, h}{}' \mathcal{F}, \quad (7.7)$$

where min' indicates minimizing only over those configurations for which the contact line at the substrate lies at a circle corresponding to the one of the reference configuration (Fig. 7.1(a)). Inserting λ from Eq. (7.5) into $F_B^\star$ yields $F_B(\alpha; f, \theta_0, \theta_p, a, R_0)$.

In both cases one can split the free energy into two parts:

$$F_\sigma(\alpha) = F_{\sigma 0} + \Delta F_\sigma(\alpha), \quad (7.8)$$

7.1. SINGLE PARTICLE

where $\sigma = A, B$. $F_{\sigma 0} := F_\sigma(\alpha = 0)$ is the free energy of the droplet with the adsorbed particle in the axially symmetric position $\alpha = 0$ and $\Delta F_\sigma := F_\sigma - F_{\sigma 0}$ is the excess free energy depending on the angular deviation of the particle position from the symmetry axis. The latter quantity will be the main focus of the following analysis.

Finally, we note that the free energy $F_{A0}(f)$ for $\alpha = 0$, as defined in Eq. (7.8), can be expressed as the Legendre transform of the free energy $\tilde{F}_{A0}(h)$ defined in Appendix E, i.e., $F_{A0}(f) = \min_h \{\tilde{F}_{A0}(h) - fh\}$.

7.1.2 Parameterization in terms of spherical coordinates and effective description of the particle

As we have shown in Chapter 5 (see, e.g., Eq. (5.19)), if without a particle the original fluid interface is flat, the capillary deformation of the interface around an added particle can be uniquely described in terms of capillary multipoles. Effectively, the interface with a particle can be replaced by the interface without the particle, but with an infinite set of point-multipoles, all centered at a single point inside the region originally occupied by the particle. In the following analysis we apply a similar approach for the case of a spherical interface. The subsequent comparison with the full numerical results will show that in the model studied here the effect of the particle onto the interface can be very well approximated by a capillary monopole. Nevertheless, for the present curved interface, we do not claim that there exists an equivalent set of point-multipoles all positioned at a single point, analogous to the case of a flat interface, which is determined uniquely by the deformation of the interface outside the particle. This actually remains an open problem.

However, one can always replace the particle by a set of capillary charges distributed over the additional virtual piece of liquid-gas interface inside the particle (not necessarily at a single point). We introduce a pressure field $\Pi(\Omega)$ defined at the virtual piece of the interface described by the solid angle $\Delta\Omega$. (In the reference configuration, this piece of interface, defined at $\Delta\Omega_{ref}$, together with the remaining part of the interface, forms a perfect cap of a sphere.) Within this approach the contact line at the particle surface is virtual and enters only through $\Pi(\Omega)$; therefore, in the following, we shall use the notion "contact line" exclusively for the contact line at the substrate. In polar coordinates, one can express the free energy functional in Eq. (7.3) as a functional of the radial deformation $u(\Omega) = r(\Omega) - R_0$:

$$\mathcal{F}[\{u(\Omega)\}] = \gamma \int_{\Omega_c} d\Omega \left[s(u, \nabla_a u) - R_0^2 \right] + \gamma R_0^2 \left(\int_{\Omega_c} - \int_{\Omega_0} \right) d\Omega$$

$$- \frac{\gamma \cos\theta_0}{2} \int_0^{2\pi} d\phi \left[(R_0 + u_c(\phi))^2 \sin^2\theta_c(\phi) - R_0^2 \sin^2\theta_0 \right]$$

$$- \frac{1}{3} \int_{\Delta\Omega} d\Omega\, \Pi(\Omega) \left[(R_0 + u)^3 - R_0^3 \right]$$

$$- \frac{\lambda}{3} \int_{\Omega_c} d\Omega \left[(R_0 + u)^3 - R_0^3 \right] + \frac{\lambda}{3} \left(\int_{\Omega_c} - \int_{\Omega_0} \right) d\Omega \left[(R_s(\theta))^3 - R_0^3 \right] + \lambda\delta V, \quad (7.9)$$

where we distinguish two integration domains:

$$\Omega_0 = \{(\theta, \phi) \in \mathbb{R}^2 \mid \phi \in [0, 2\pi) \wedge \theta \in [0, \theta_0]\}, \quad (7.10)$$

which corresponds to the reference droplet shape, and Ω_c in which the actual droplet interface $u(\Omega)$ is defined:

$$\Omega_c = \{(\theta, \phi) \in \mathbb{R}^2 \mid \phi \in [0, 2\pi) \wedge \theta \in [0, \theta_c(\phi)]\}. \tag{7.11}$$

Ω_c differs from Ω_0 only by the maximal polar angle $\theta_c = \theta_c(\phi)$, which determines the shape of the contact line. Equivalently, the shape of the contact line is also described by the deformation $u_c = u_c(\phi) \equiv u(\theta = \theta_c(\phi), \phi)$. In the case of a pinned contact line one has $\Omega_c = \Omega_0$ and $u_c = 0$, but in the case of a free contact line in general $\Omega_c \neq \Omega_0$. We note that the parameterization in terms of spherical coordinates remains valid for $\theta_0 > \pi/2$ (in this case the center of the droplet lies above the substrate). The first two terms on the rhs of Eq. (7.9) emerge from $\gamma [\int_{\Omega_c} d\Omega\, s - \int_{\Omega_0} d\Omega\, s_{ref}]$ with $s_{ref} = R_0^2$ and thus they represent the changes in the surface energy of the liquid-gas interface. The third term represents the changes in the liquid-substrate surface energy. The fourth term represents the work done by the external pressure $\Pi(\Omega)$ in displacing the interface and the last three terms correspond to the volume conservation. The last but one term corrects the previous one with a liquid volume wedged between the substrate surface and the surface of the cap of the reference sphere inside the domains $\Omega_c \setminus \Omega_0$ and $\Omega_0 \setminus \Omega_c$ (in the former case the reference surface is extended into the domain $\Omega_c \setminus \Omega_0$). Accordingly, $R_s(\theta) = R_0 \cos\theta_0 / \cos\theta$ expresses the distance between points on the substrate surface and the origin in terms of spherical coordinates. We note that in this term the brackets do not delimit the action of the integral. The last term represents a correction δV corresponding to the change of the volume (due to an interface displacement u) of the virtual liquid domain added inside the particle (see Fig. 7.2). Comparing the expressions in Eqs. (7.9) and (7.3) yields an implicit definition of the effective pressure field Π, in the sense that $\Pi(\Omega)$ must be chosen such that the following equation is fulfilled:

$$\frac{1}{3}\int_{\Delta\Omega} d\Omega\, \Pi(\Omega)\Big[(R_0+u)^3 - R_0^3\Big] = fh + \gamma \delta S_{pl} \cos\theta_p + \gamma \delta S_{lg}, \tag{7.12}$$

where $\delta S_{pl} = S_{pl} - S_{pl,ref}$; δS_{lg} is the change of the area of the virtual liquid-gas interface relative to the reference configuration (see Fig. 7.2). Equation (7.12) does not uniquely determine $\Pi(\Omega)$ but in the limit of small particles, as will be studied in Subsec. 7.2.4, it provides a condition sufficient for calculating the shape of the droplet and thus serves as an implicit definition of the pressure Π. In the present notation Π has actually the meaning of Π_{ext} introduced in Sec. 6.3, but for simplicity, in the following, we omit the subscript ext.

The shift of the center of mass is given by

$$\begin{aligned} x_{CM} - x_{CM,ref} &= \frac{1}{V_l}\left(\int_{\Omega_c} d\Omega \int_{R_s(\theta)}^{R_0+u} dr - \int_{\Omega_0} d\Omega \int_{R_s(\theta)}^{R_0} dr\right) r^3 \sin\theta \cos\phi \;-\; \delta x \\ &= \frac{1}{4V_l}\int_{\Omega_c} d\Omega \left[(R_0+u)^4 - R_0^4\right] \sin\theta \cos\phi \\ &\quad - \frac{1}{4V_l}\left(\int_{\Omega_c} - \int_{\Omega_0}\right) d\Omega \left[(R_s(\theta))^4 - R_0^4\right]\sin\theta \cos\phi \;-\; \delta x, \end{aligned} \tag{7.13}$$

7.2. PERTURBATION THEORY

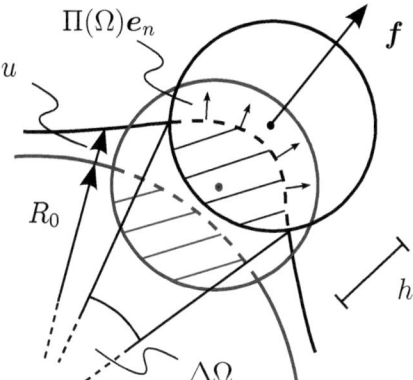

Figure 7.2: Schematic cross-sectional representation of the virtual piece of the liquid-gas interface (dashed lines) added inside the particle on which $\Pi(\Omega)$ is defined (e_n is a unit vector normal to that interface). The quantity δS_{lg} (see main text) is the change in the surface area of that piece relative to the reference configuration ($u = 0$), whereas δV is the volume change of the corresponding virtual liquid domain inside the particle (hatched regions). The displacement of the center of the particle is denoted as h. The radial displacement of the interface relative to the reference configuration (R_0) is denoted as u.

where the second term in the final expression represents the contribution to the shift due to the deformation of the contact line and δx is the contribution to the shift of the center of mass of the whole droplet due to the shift of the virtual liquid domain assigned to the inside of the particle. $R_s(\theta)$ has the same meaning as in Eq. (7.9).

7.2 Perturbation theory

It is the aim of this section to base our analytical approach on a systematic expansion in terms of a small parameter ϵ, such that the deformations of the droplet are small, too.

We first note that strongly deformed droplets are usually metastable or unstable (see Appendix E), so that a necessary condition for the deformations being small is the stability of the droplet shape. In the case of a spherical droplet this means that the shift in the internal pressure due to the force f acting on the interface must be much smaller than the Laplace pressure $2\gamma/R_0$. Up to a geometrical factor the shift is of the order of f/R_0^2, so that in order to avoid such instabilities one must have

$$|f| \ll \gamma R_0, \tag{7.14}$$

which means that the external force must be much smaller than the maximal capillary force acting on the contact line at the substrate which is of the order of γR_0. On the other hand the same reasoning applies to the contact line at the particle implying that the force f should not be stronger than the maximal capillary force acting on the particle which is of the order of γa. One can argue that even if the deformation of the interface is large (in a not yet precisely defined sense) near the particle, perturbation theory could still be valid far away from the particle. On the other hand one can expect that the long-ranged behavior of the deformation field around the particle actually determines the free energy landscape ΔF_σ (unless the particle is very close to the substrate). Thus, for the purpose of calculating ΔF_σ, instead of $|f| \ll \gamma a$ one only needs

$$|f| \lesssim \gamma a. \tag{7.15}$$

Therefore, it is reasonable to choose

$$\epsilon = \frac{|f|}{\gamma R_0} \tag{7.16}$$

as a small dimensionless parameter, assuming that the condition in Eq. (7.15) is fulfilled, too. We note that for small particles, i.e., for

$$a \ll R_0, \tag{7.17}$$

the condition $\epsilon \ll 1$ is automatically satisfied. In such a case the expansion in terms of ϵ also leads to the asymptotic result $F_{\sigma 0} \sim f^2 \ln(R_0/a)$ for the self-energy (see Eq. (7.59)), which reproduces the behavior $F_{\sigma 0} \sim f^2 \ln(\lambda_c/a)$ for a flat interface with $\lambda_c = R_0$ being the capillary length. In the limit of a small particle, i.e., $a \to 0$, with $f/(\gamma a) = const$, the following reasoning shows that the region of strong interfacial gradients in the neighborhood of the particle actually contributes only to the subleading term $O(a^2)$ in the free energy. First, the interfacial gradients become $O(1)$ only on the scale of a, because for small deformations u of a locally flat interface one has $\nabla_\| u \approx f/(2\pi\gamma r)$, where r is the distance from the particle. Then, due to the stability condition $|f| \lesssim \gamma a$ (Eq. (7.15)) one has $|\nabla_\| u| \sim 1$ only for $r \sim a$. Nevertheless, one can still apply the linear theory by introducing a notion of an effective colloidal particle which incorporates all the region of strong interfacial gradients. From the above reasoning it follows that the size of this effective particle is of the order of a. Therefore the free energy corresponding to this region is $\sim a^2$, i.e., it contributes only to the subleading term compared to the leading one $\sim a^2 \ln(R_0/a)$ (where we assume $f \sim a$). Finally, in the case of large particles, such that $a \lesssim R_0$, one should rather use $|f|/(\gamma a)$ as a small parameter.

7.2.1 Systematic expansion, variation, boundary conditions, and force balance

In the next step, we expand u, λ, Π, f_{CM}, and, for later purposes, h in terms of the small parameter ϵ:

$$u(\Omega) = R_0\big(\epsilon\tilde{v}(\Omega) + O(\epsilon^2)\big), \tag{7.18}$$

$$\lambda = \frac{\gamma}{R_0}\big(2 + \epsilon\mu + O(\epsilon^2)\big), \tag{7.19}$$

$$\Pi(\Omega) = \frac{\gamma}{R_0}\big(\epsilon\pi(\Omega) + O(\epsilon^2)\big), \tag{7.20}$$

$$f_{CM} = \gamma R_0\big(\epsilon Q_{CM} + O(\epsilon^2)\big), \tag{7.21}$$

$$h = R_0\big(\epsilon\tilde{h} + O(\epsilon^2)\big), \tag{7.22}$$

where λ has been expanded around the Laplace pressure $2\gamma/R_0$ of a spherical droplet; $\tilde{v}$, μ, π, Q_{CM}, and $\tilde{h}$ are dimensionless. In the following we shall drop the tilde above v, i.e., $\tilde{v} \equiv v$.

In the following we shall keep the radius R_0 of the droplet fixed and therefore it is convenient to divide the energy by γR_0^2. In zeroth order in ϵ the droplet is undeformed ($v \equiv 0$) and the free energy functional $\mathcal{F}^{(0)}$ as defined in Eq. (7.9) equals zero. Similarly, the free energy $\mathcal{F}^{(1)}$ in first order in ϵ also vanishes, which reflects the fact that we have chosen the equilibrium configuration as the reference one (in equilibrium, by definition, perturbations do not give rise to linear contributions to the free energy). Therefore the leading contribution is of second order in ϵ and the free energy functional reads (see Appendix B)

$$\frac{\mathcal{F}}{\gamma R_0^2} = \epsilon^2 \int_{\Omega_0} d\Omega \left[\frac{1}{2}(\nabla_a v)^2 - v^2 - \big(\pi(\Omega) + \mu\big)v\right] - \frac{\epsilon^2}{2}\cos\theta_0 \int_0^{2\pi} d\phi \, (v|_{\theta_0})^2 + O(\epsilon^3), \tag{7.23}$$

where $\pi(\Omega \notin \Delta\Omega) = 0$ and the boundary term (second term) has been obtained by expanding up to second order in ϵ those terms in the free energy functional in Eq. (7.9), which depend on the deformations of the contact line. (For $\theta_c(\phi) < \theta_0$, the deformation $v(\theta)$ is smoothly extended to $\theta = \theta_0$, so that in Eq. (7.23) $v(\theta)$ is defined within the whole angular domain Ω_0; this is justified because the corresponding difference in the free energy is of higher order $O(\epsilon^3)$.) Finally, we note that the general form of the free energy functional in Eq. (7.23) has been already derived in covariant form by Rosso & Virga (2003) (see Eq. (11) therein) and also by Brinkmann et al. (2004) (see Eq. (82) therein) for the purpose of the stability analysis of liquid drops placed on arbitrarily shaped substrates including a non-vanishing line tension. As can be checked, in the special case of a flat substrate and a vanishing line tension those results reduce to Eq. (7.23).

The variation of $\mathcal{F}$ renders

$$\frac{1}{\gamma R_0^2}\left(\mathcal{F}[v+\delta v] - \mathcal{F}[v]\right) = \epsilon^2 \int_{\Omega_0} d\Omega\,[-\nabla_a^2 v - 2v - \pi(\Omega) - \mu]\delta v + \epsilon^2 \int_{\Omega_0} d\Omega\, \nabla_a\cdot(\delta v \nabla_a v)$$
$$- \epsilon^2 \cos\theta_0 \int_0^{2\pi} d\phi\, v|_{\theta_0}\delta v|_{\theta_0} + O((\delta v)^2, \epsilon^3), \quad (7.24)$$

whereas the variation of the additional term responsible for fixing the center of mass (Eq. (7.13)) yields

$$-\frac{f_{CM}}{\gamma R_0^2}(x_{CM} - x_{CM,ref})|_v^{v+\delta v} = -\epsilon^2 \frac{3Q_{CM}}{4\pi f_0(\theta_0)} \int_{\Omega_0} d\Omega\, \delta v \sin\theta \cos\phi + O((\delta v)^2, \epsilon^3), \quad (7.25)$$

where we have neglected the second term in Eq. (7.13), which is of the order $O(\epsilon^2)$ and thus in Eq. (7.25) it contributes only to the order $O(\epsilon^3)$, and δx which contributes only to the order $O(\epsilon^2) \times O(a/R_0)^3$. The function $f_0(\theta_0)$ expresses the volume of a unit spherical cap and it has been defined in Eq. (E.33).

The expressions in Eqs. (7.24) and (7.25) lead to the Euler-Lagrange equation for a droplet with a fixed center of mass in the form

$$-(\nabla_a^2 + 2)v(\Omega) = \pi(\Omega) + \pi_{CM}(\Omega) + \mu, \quad (7.26)$$

where

$$\pi_{CM}(\Omega) = \frac{3Q_{CM}}{4\pi f_0(\theta_0)} \sin\theta \cos\phi \quad (7.27)$$

is an effective pressure fixing the center of mass of the droplet. In the case that the center of mass is not fixed, the corresponding Euler-Lagrange equation has the same form as in Eq. (7.26) but with $\pi_{CM} = 0$. Applying the divergence theorem (on a unit sphere) to the second term in Eq. (7.24) gives the boundary contribution

$$\frac{1}{\gamma R_0^2}\left(\mathcal{F}[v+\delta v] - \mathcal{F}[v]\right)_{bc} = \epsilon^2 \int_0^{2\pi} d\phi\,(\sin\theta_0 \partial_\theta v|_{\theta_0} - \cos\theta_0 v|_{\theta_0})\delta v|_{\theta_0}, \quad (7.28)$$

vanishing if $\delta v|_{\theta_0} = 0$, which corresponds to a pinned contact line (model B). If the contact line is free (model A) one has $\delta v|_{\theta_0} \neq 0$ and, instead, the expression in brackets of the integrand in Eq. (7.28) must vanish. These two different conditions correspond to Dirichlet (B) and Robin (A) boundary conditions, respectively:

$$A\ (\text{free}): \qquad \sin\theta_0 \partial_\theta v|_{\theta_0} - \cos\theta_0 v|_{\theta_0} = 0, \quad (7.29)$$
$$B\ (\text{pinned}): \qquad v|_{\theta_0} = 0. \quad (7.30)$$

Equation (7.29) is equivalent to the condition that the contact angle is equal to the Young angle, which can be seen by the following reasoning. In the limit of small deformations the normal $\boldsymbol{e}_n$ to the droplet surface in terms of spherical coordinates can be expressed as $\boldsymbol{e}_n = \boldsymbol{e}_r - \epsilon \nabla_a v + O(\epsilon^2)$, so that the contact angle $\tilde{\theta}(\phi)$ at the substrate is given by

$$\cos\tilde{\theta}(\phi) = \boldsymbol{e}_z \cdot \boldsymbol{e}_n|_{\theta_c} = \cos\theta_c + \epsilon \sin\theta_c \partial_\theta v|_{\theta_c} + O(\epsilon^2) =$$
$$= \cos\theta_0 + \epsilon[\sin\theta_0 \partial_\theta v|_{\theta_0} - v|_{\theta_0}\cos\theta_0] + O(\epsilon^2), \quad (7.31)$$

7.2. PERTURBATION THEORY

where in the third equality we used $\cos\theta_c = \cos\theta_0(1 - \epsilon v|_{\theta_0}) + O(\epsilon^2)$, which follows from the analysis of small perturbations of a spherical cap. Thus, from Eqs. (7.29) and (7.31) we recover Young's law in the form $\bar{\theta}(\phi) = \theta_0$. We note that whereas the condition in Eq. (7.30) corresponds to the shape of the droplet at the actual contact line, because for a pinned contact line $\theta_c(\phi) = \theta_0$, the condition in Eq. (7.29) does not apply directly at the contact line $\partial\Omega_c$ but at the boundary $\partial\Omega_0$ of the reference integration domain. The actual shape of the contact line in this case can be obtained by a linear extrapolation of the interface profile from θ_0 to θ_c.

The volume constraint in Eq. (7.5) in first order in ϵ reads

$$\int_{\Omega_0} d\Omega\, v(\Omega) = 0, \tag{7.32}$$

whereas in Eq. (7.6) the constraint of fixed center of mass, due to Eqs. (7.13) and (7.27), can be written, up to second order in ϵ, as

$$\int_{\Omega_0} d\Omega\, \pi_{CM}(\Omega) v(\Omega) = 0. \tag{7.33}$$

In Eqs. (7.32) and (7.33) we have neglected contributions from δV (Eq. (7.9)) and δx (Eq. (7.13)), which are both of the order $O(a^2/R_0^2)$ and $O(a^3/R_0^3)$, respectively.

In the following our goal is to express the pressure shift μ and the balance of forces acting on the droplet in terms of π and θ_0. Integrating both sides of Eq. (7.26) over Ω_0, with $\int_{\Omega_0} d\Omega = 2\pi(1 - \cos\theta_0)$, and applying the divergence theorem and subsequently the condition of constant volume (Eq. (7.32)) yields

$$\mu = -\frac{1}{2\pi(1 - \cos\theta_0)} \left[\int_{\Omega_0} d\Omega\, \pi(\Omega) + \sin\theta_0 \int_0^{2\pi} d\phi\, \partial_\theta v|_{\theta_0} \right], \tag{7.34}$$

where we have used the fact that $\int_{\Omega_0} d\Omega\, \pi_{CM} = 0$. Equation (7.34) can be interpreted as a hydrostatic balance in that the internal pressure (lhs) is equal to the external pressure exerted by the forces acting at the droplet (rhs), which in particular depend on the boundary conditions.

The governing equation (7.26) can also be used to derive the force balance on the droplet. First, multiplying both sides by the radial vector $\mathbf{e}_r(\Omega)$ and integrating over Ω_0 we obtain

$$-\int_{\Omega_0} d\Omega\, \mathbf{e}_r(\Omega)(\nabla_a^2 + 2)v = \int_{\Omega_0} d\Omega\, \pi(\Omega)\mathbf{e}_r(\Omega) + Q_{CM}\mathbf{e}_x + \mu\pi\sin^2\theta_0 \mathbf{e}_z. \tag{7.35}$$

The first term on the rhs of Eq. (7.35) represents, up to first order in ϵ, the total external force acting on the droplet. The surface integral on the lhs, after integrating by parts, using the fact that $(\nabla_a^2+2)\mathbf{e}_r(\Omega) = 0$, which follows from the identity $(\nabla_a^2+2)Y_{1m} = 0$, and finally applying the divergence theorem can be transformed into a line integral,

$$-\int_{\Omega_0} d\Omega\, \mathbf{e}_r(\Omega)(\nabla_a^2 + 2)v = -\sin\theta_0 \int_0^{2\pi} d\phi\, (\mathbf{e}_r \partial_\theta v - v\partial_\theta \mathbf{e}_r)|_{\theta_0}. \tag{7.36}$$

Using $\partial_\theta \mathbf{e}_r = \mathbf{e}_\theta$ and taking the limit $a/R_0 \to 0$ Eq. (7.35) can be expressed in terms of Cartesian coordinates:

$$x: \quad -\sin\theta_0 \int_0^{2\pi} d\phi \, \cos\phi \Big(\sin\theta_0 \partial_\theta v|_{\theta_0} - v|_{\theta_0} \cos\theta_0 \Big) = \int_{\Omega_0} d\Omega \, \pi(\Omega) \sin\theta \cos\phi + Q_{CM}, \tag{7.37}$$

$$y: \quad -\sin\theta_0 \int_0^{2\pi} d\phi \, \sin\phi \Big(\sin\theta_0 \partial_\theta v|_{\theta_0} - v|_{\theta_0} \cos\theta_0 \Big) = 0, \tag{7.38}$$

$$z: \quad -\sin\theta_0 \int_0^{2\pi} d\phi \, \Big(\cos\theta_0 \partial_\theta v|_{\theta_0} + v|_{\theta_0} \sin\theta_0 \Big) = \int_{\Omega_0} d\Omega \, \pi(\Omega) \cos\theta + \mu\pi \sin^2\theta_0. \tag{7.39}$$

In the case of a free contact line (see Eq. (7.29)) the left hand sides of Eqs. (7.37) and (7.38) vanish and the force balance in the x-direction reduces to

$$-Q_{CM} = \int_{\Omega_0} d\Omega \, \pi(\Omega) \sin\theta \cos\phi. \tag{7.40}$$

If $\pi(\Omega)$ corresponds to a pointlike particle (see, c.f., Sec. 7.2.4), so that $\pi(\Omega) = Q\delta(\Omega, \Omega_1) = Q\delta(\theta - \theta_1)\delta(\phi - \phi_1)/\sin\theta$, where $Q = f/|f|$ and $\Omega_1 = (\theta_1 = \alpha, \phi_1 = 0)$ is the direction along which the external force pulls ($Q = +1$) or pushes ($Q = -1$) the particle, due to $f_{CM} = \epsilon\gamma R_0 Q_{CM} = |f|Q_{CM}$ (see Eqs. (7.16) and (7.21)) Eq. (7.40) leads to $f_{CM} = -f \sin\alpha$. Accordingly, the Lagrange multiplier f_{CM} (Eq. (7.4)) can indeed be interpreted as a force counterbalancing the x-component of the external force, equal to $f \sin\alpha$, and thus fixing the center of mass of the droplet. In the same limiting case the first term on the rhs of Eq. (7.35) reduces to $\int_{\Omega_0} d\Omega \, \pi(\Omega) \mathbf{e}_r(\Omega) = q\mathbf{e}_r(\Omega_1)$ corresponding to the external force on the droplet. In the case of a pinned contact line and with a free center of mass (i.e., $f_{CM} = 0$ so that $Q_{CM} = 0$) the force balance in the x-direction (Eq. (7.37)) reads

$$-\sin^2\theta_0 \int_0^{2\pi} d\phi \, \cos\phi \, \partial_\theta v|_{\theta_0} = \int_{\Omega_0} d\Omega \, \pi(\Omega) \sin\theta \cos\phi. \tag{7.41}$$

Thus, in this case, the x-component of the external force (rhs) is counterbalanced by the capillary force due to the deformation of the droplet at the contact line (lhs).

For model A the lhs of Eq. (7.38) vanishes due to Eq. (7.29), but it also vanishes for model B because $\partial_\theta v|_{\theta_0}$ and $v|_{\theta_0}$ are symmetric and $\sin\phi$ is antisymmetric with respect to the xz-plane.

The difference between these two models is that for model A the local contact angle is the Young angle (see Eqs. (7.31) and (7.29)), which guarantees that each piece of the contact line is in mechanical equilibrium, whereas in the case of model B this is not the case and the total capillary force on the contact line counterbalances the x-component of the external force (Eq. (7.41)).

The boundary condition in Eq. (7.29) allows one to express $\partial_\theta v|_{\theta_0}$ in terms of $v|_{\theta_0}$. Inserting this into Eqs. (7.34) and (7.39) renders two equations for μ and $\int_0^{2\pi} d\phi \, v|_{\theta_0}$,

7.2. PERTURBATION THEORY

which can be solved to yield

$$A: \qquad \mu = -\frac{1}{4\pi f_0(\theta_0)} \int_{\Omega_0} d\Omega\, \pi(\Omega)(1 - \cos\theta_0 \cos\theta), \qquad (7.42)$$

$$B: \qquad \mu = -\frac{1}{\pi(1 - \cos\theta_0)^2} \int_{\Omega_0} d\Omega\, \pi(\Omega)(\cos\theta - \cos\theta_0). \qquad (7.43)$$

Equation (7.43) follows from inserting the boundary condition in Eq. (7.30) into Eq. (7.39); this renders $\int_0^{2\pi} d\phi\, \partial_\theta v|_{\theta_0}$ in terms of μ, which can be inserted into Eq. (7.34) yielding an equation for μ. In the case of a pointlike particle (see, c.f., Eq. (7.58)), Eqs. (7.42) and (7.43) reduce to $\mu = -Q(1 - \cos\theta_0 \cos\alpha)/(4\pi f_0(\theta_0))$ and $\mu = -Q(\cos\alpha - \cos\theta_0)/(\pi(1-\cos\theta_0)^2)$, respectively, so that due to $\alpha < \theta_0$ (see Fig. 7.1(a)), the internal uniform pressure shift μ has always the sign $-Q$, i.e., the opposite one to f. This means that the pressure change counteracts the action of the external force, which can be interpreted as the realization of Le Chatelier's principle for this particular system.

7.2.2 Green's functions

In this subsection we present a formal solution of Eq. (7.26) by means of Green's functions $G_\sigma(\Omega, \Omega')$ which are the radial deformations $v(\Omega)$ due to a pointlike perturbation $\pi(\Omega) = \delta(\Omega, \Omega')$ acting in the direction $\Omega' = (\theta', \phi')$. Accordingly the Green's functions fulfill Eq. (7.26) with $\pi_{CM} = 0$ for model B, π_{CM} given by Eqs. (7.27) and (7.40) for model A, and μ given by Eqs. (7.42) and (7.43). Note that Eq. (7.27) corresponds to the special case that the external force acts in the xz-plane, i.e., $\phi' = 0$ so that in the general case $\cos\phi$ in Eq. (7.27) is replaced by $\cos(\phi - \phi')$. As a consequence the Green's functions fulfill

$$-(\nabla_a^2 + 2)G_A(\Omega, \Omega') = \delta(\Omega, \Omega') - \frac{3}{4\pi f_0(\theta_0)} \sin\theta' \sin\theta \cos(\phi - \phi') - \frac{1 - \cos\theta_0 \cos\theta'}{4\pi f_0(\theta_0)} \qquad (7.44)$$

for model A and

$$-(\nabla_a^2 + 2)G_B(\Omega, \Omega') = \delta(\Omega, \Omega') - \frac{\cos\theta' - \cos\theta_0}{\pi(1 - \cos\theta_0)^2} \qquad (7.45)$$

for model B, where in both cases $\Omega, \Omega' \in \Omega_0$. We note that for arbitrary θ_0 the Green's functions are not symmetric, i.e., $G_\sigma(\Omega, \Omega') \neq G_\sigma(\Omega', \Omega)$. This means that the deformation at Ω due to a perturbation at Ω' in general differs from the deformation at Ω' due to a perturbation at Ω. This can be understood by the fact that the shift μ of the internal pressure depends on where the perturbation is applied. The only exception is the case $\theta_0 = \pi/2$ with a free contact line, for which the Green's function G_A is fully symmetric. In this case $\cos\theta_0 = 0$ so that the last term on the rhs of Eq. (7.44) is a constant, whereas the last but one term is explicitly symmetric.

The deformation of the droplet for an arbitrary pressure field $\pi(\Omega)$ can be expressed in terms of the Green's functions:

$$v_\sigma(\Omega) = \int_{\Omega_0} d\Omega'\, G_\sigma(\Omega, \Omega') \pi(\Omega'). \qquad (7.46)$$

By using Eqs. (7.44) and (7.45) one can check that this expression for v_σ fulfills the Young-Laplace equation (7.26). For $\pi(\Omega) = \delta(\Omega, \Omega')$ the volume constraint in Eq. (7.32) yields

$$\int_{\Omega_0} d\Omega\, G_\sigma(\Omega, \Omega') = 0, \qquad (7.47)$$

whereas the boundary conditions in Eqs. (7.29) and (7.30) can be expressed as

$$\sin\theta_0 \partial_\theta G_A(\Omega, \Omega')|_{\Omega \in \partial\Omega_0} - \cos\theta_0 G_A(\Omega, \Omega')|_{\Omega \in \partial\Omega_0} = 0, \qquad (7.48)$$

$$G_B(\Omega, \Omega')|_{\Omega \in \partial\Omega_0} = 0. \qquad (7.49)$$

7.2.3 Free energy

In this Subsection we derive the expression for the free energy of a sessile droplet in mechanical equilibrium and subjected to a pressure field $\pi(\Omega)$ (Eq. (7.51)). In terms of perturbation theory, the free energy both for model A and model B is given by the functional in Eq. (7.23) evaluated for v obeying the Young-Laplace equation (7.26) with $\pi_{CM} = 0$ for model B and with the condition in Eq. (7.33) for model A. This yields

$$\begin{aligned}F_\sigma &= \frac{f^2}{\gamma} \int_{\Omega_0} d\Omega \left[\frac{1}{2}\nabla_a(v\nabla_a v) - \frac{1}{2}v(\nabla_a^2 v + 2v) - (\pi(\Omega) + \mu)v\right]\\ &\quad - \frac{f^2}{2\gamma}\cos\theta_0 \int_0^{2\pi} d\phi\, (v|_{\theta_0})^2 \\ &= -\frac{f^2}{2\gamma}\int_{\Omega_0} d\Omega\, \pi(\Omega) v + \frac{f^2}{2\gamma}\int_0^{2\pi} d\phi\, v|_{\theta_0}\left(\sin\theta_0 \partial_\theta v|_{\theta_0} - \cos\theta_0 v|_{\theta_0}\right),\end{aligned} \qquad (7.50)$$

where the second equality follows from applying the divergence theorem, Eq. (7.26), Eq. (7.33), and the constant volume constraint (Eq. (7.32)). The second term in the last expression vanishes for boundary conditions corresponding to either a free or a pinned contact line at the substrate (see Eqs. (7.29) and (7.30)). This leads to the free energy in the form

$$F_\sigma = -\frac{1}{2}\epsilon^2 \gamma R_0^2 \int_{\Omega_0} d\Omega\, \pi(\Omega) v(\Omega) = -\frac{1}{2}\epsilon^2 \gamma R_0^2 \int_{\Omega_0} d\Omega \int_{\Omega_0} d\Omega'\, \pi(\Omega) G_\sigma(\Omega, \Omega') \pi(\Omega'), \qquad (7.51)$$

where for the second equality we have used the general form of the solution given in Eq. (7.46).

7.2.4 Limit of small particles

Taking the limit $a \to 0$ allows one to provide a relation between the unknown pressure field π and the system parameters introduced in Subsec. 7.1.1. In this context, Eq. (7.12) provides a constraint on π expressed in terms of an unknown deformation field, encoded by the quantities h, δS_{pl}, and δS_{lg}. However, this equation simplifies considerably in the limit of small particles. In order to see this, we consider Eqs. (7.44) and (7.45) in the limit $\Omega \to \Omega'$, i.e., when the spherical reference interface becomes

7.2. PERTURBATION THEORY

locally flat in the neighborhood of Ω and Ω'. First, we rotate the coordinate frame such that the z-axis points in an arbitrarily chosen direction in the neighborhood of Ω and Ω' (in the following we refer to this new axis as $\tilde{z}$-axis). Next, we project Ω and Ω' along the $\tilde{z}$-axis onto the plane tangent to the reference spherical interface at the point of intersection of this interface with the $\tilde{z}$-axis. The mapping $(\tilde{\theta}, \tilde{\phi}) \mapsto (\tilde{\rho}, \tilde{\phi})$, where $(\tilde{\theta}, \tilde{\phi})$ denote the spherical coordinates associated with the $\tilde{z}$-axis and $(\tilde{\rho}, \tilde{\phi})$ are polar coordinates in the tangent plane, is given by

$$\tilde{\rho} = R_0 \sin \tilde{\theta}. \qquad (7.52)$$

For simplicity of the notation we skip the tilde so that $(\tilde{\rho}, \tilde{\phi}) \equiv (\rho, \phi)$. According to the analysis performed in Appendix C, Eqs. (7.44) and (7.45) can be rewritten in terms of the new coordinates in the form of a single equation:

$$-(R_0^2 \nabla_\parallel^2 + 2 - \partial_\rho \rho^2 \partial_\rho) G_\sigma(\boldsymbol{x}, \boldsymbol{x}') = R_0 \sqrt{R_0^2 - \rho^2}\, \delta(\boldsymbol{x} - \boldsymbol{x}') + \Delta_\sigma(\boldsymbol{x}, \boldsymbol{x}'), \qquad (7.53)$$

where $\nabla_\parallel^2$ is the Laplace operator on the tangent plane and $G_\sigma(\boldsymbol{x}, \boldsymbol{x}') := G_\sigma(\Omega(\boldsymbol{x}), \Omega'(\boldsymbol{x}'))$; the function $\Delta_\sigma(\boldsymbol{x}, \boldsymbol{x}') := \Delta_\sigma(\Omega(\boldsymbol{x}), \Omega'(\boldsymbol{x}'))$ denotes all the regular non-homogeneous terms on the right hand sides of Eqs. (7.44) and (7.45). Applying the limit $R_0 \to \infty$ to Eq. (7.53) leads to the usual two-dimensional Green's equation

$$-\nabla_\parallel^2 G_\sigma(\boldsymbol{x}, \boldsymbol{x}') = \delta(\boldsymbol{x} - \boldsymbol{x}'), \qquad (7.54)$$

with the solution $G_\sigma(\boldsymbol{x}, \boldsymbol{x}') \equiv G(\boldsymbol{x}, \boldsymbol{x}') = G(|\boldsymbol{x} - \boldsymbol{x}'|) = -(1/2\pi) \ln(|\boldsymbol{x} - \boldsymbol{x}'|/R_0)$ (compare with Eqs. (5.12) and (5.13); here R_0 is a natural choice of the large distance cut-off), which depends neither on σ nor on $\boldsymbol{x}'$, or equivalently Ω', i.e., on where on the droplet the perturbation is applied.

Therefore the deformation v_σ diverges logarithmically in the neighborhood of a pointlike perturbation at Ω' independently of the boundary conditions far away from Ω'. By taking $\pi(\Omega) = \delta(\Omega, \Omega')$ in Eq. (7.46) one obtains

$$v_\sigma(\Omega) = G_\sigma(\Omega, \Omega') \xrightarrow[\Omega \to \Omega']{} -\frac{1}{2\pi} \ln(\bar{\theta}). \qquad (7.55)$$

where $\bar{\theta}$ is the angle between the unit vectors pointing into the directions Ω and Ω'. Thus, with the radius a of the particle acting as a natural cutoff, the dimensionless displacement $\tilde{h}$ of the particle (see Eq. (7.22)) in the leading order in a/R_0 reads

$$\tilde{h} = \frac{Q}{2\pi} \ln\left(\frac{R_0}{a}\right) + O(a/R_0), \qquad (7.56)$$

where $Q = f/|f| = \mathrm{sgn}(f)$. Accordingly Eq. (7.12) in leading order in ϵ can be written as (see Eqs. (7.18), (7.20), (7.22), and (7.16))

$$\int_{\Omega_0} d\Omega\, \pi(\Omega) v(\Omega) = \frac{1}{2\pi} \ln\left(\frac{R_0}{a}\right) + O(1). \qquad (7.57)$$

The correction term in Eq. (7.57), which stems from the last two terms on the rhs of Eq. (7.12), is proportional to $\gamma^2 a^2/f^2$ which, in agreement with Eq. (7.15), is of order 1 relative to the leading term which diverges $\sim \ln(R_0/a)$. For $f \to 0$ the result is the same because in this case the last two terms on the rhs of Eq. (7.12) add up to $O(f^2)$, which leads to a correction $O(f^2/f^2) = O(1)$ in Eq. (7.57). This follows from the fact that one considers a perturbation of a stationary point of the free energy functional for which the contributions $O(f)$ must vanish. This result can also be obtained explicitly by expanding δS_{lg} and δS_{pl} in terms of $\epsilon_F := f/(2\pi\gamma a)$ analogously to the expansion performed in Appendix A in Oettel et al. (2005b) (but here for the region inside the particle instead of outside). Furthermore, assuming that the virtual piece of the interface defined at $\Delta\Omega$ is bounded to lie inside the particle, one has $u(\Omega)|_{\Delta\Omega} = h + O(a/R_0)$ (see Fig. 7.2) so that $v(\Omega)|_{\Delta\Omega} = \tilde{h} + O(a/R_0)$. According to Eqs. (7.56) and (7.57) this leads to

$$\int_{\Omega_0} d\Omega\, \pi(\Omega) = Q + O(1/\ln(R_0/a)). \tag{7.58}$$

Equations (7.57) and (7.58) are satisfied by $\pi(\Omega) = Q\,\delta(\Omega, \Omega_1)$ with $\Omega_1 = (\theta = \alpha, \phi = 0)$ (see Fig. 7.1). Thus, in leading order in a/R_0 the integrated amplitude of the effective pressure is independent of the particle radius a. In this limit the droplet radius R_0 is the only remaining length scale.

In Eq. (7.51) it is useful to consider the singular contribution to $G_\sigma(\Omega, \Omega')$ separately by writing $G_\sigma = G + G_{\sigma,reg}$, such that the regular part $G_{\sigma,reg} = G_\sigma - G$ does not diverge for $\Omega \to \Omega'$. (As described in the subsequent section in special cases G can be identified as Green's function for a free droplet (see, c.f., Eq. (6.26)).)

With $\pi(\Omega) = Q\,\delta(\Omega, \Omega_1)$ this leads to the following expressions for the free energy:

$$F_{\sigma 0} = -\frac{f^2}{2\gamma}\Big[G(\bar{\theta} = a/R_0) + G_{\sigma,reg}(0,0) + O(1)\Big] \tag{7.59}$$

and

$$\Delta F_\sigma = -\frac{f^2}{2\gamma}\Big[G_{\sigma,reg}(\Omega_1, \Omega_1) - G_{\sigma,reg}(0,0) + O(a/R_0)\Big]. \tag{7.60}$$

where $G(\Omega_1, \Omega_1)$ for $\Omega_1 = 0$ (the apex position) has been regularized as $G(\bar{\theta} = a/R_0)$. The leading term of the free energy $F_{\sigma 0}$ stems from the singular part of the kernel and thus, in analogy to electrostatics, it can be interpreted as a *self-energy* of the particle. However, unlike the self-energy of a point-charge in electrostatics, this quantity does not diverge because the singularity is suppressed by the prefactor f^2 (Eq. (7.59)) the upper bound of which, as already noted, is proportional to $\gamma^2 a^2$ (Eq. (7.15)). Thus for $a \to 0$ with $f/(\gamma a)$ kept constant the self-energy $F_{\sigma 0}$ vanishes as $a^2 \ln(R_0/a)$. In the following section we shall rather focus on the remaining part of the free energy ΔF_σ which determines the angular free energy landscape of the particle. This quantity is well-defined even in the limiting case that $\pi(\Omega)$ is of the form $\pi(\Omega) = Q\,\delta(\Omega, \Omega_1)$. In this case one obtains the closed expression (Eqs. (7.60), (7.64) and (7.67))

$$\Delta F_\sigma(\alpha) = -\frac{f^2}{2\gamma}[g_\sigma(\alpha) - g_\sigma(0)], \tag{7.61}$$

7.3. METHOD OF IMAGES FOR $\theta_0 = \pi/2$ 99

where the functions g_A and g_B are both regular at $\alpha = 0$ and also both depend on θ_0. As will be shown in the next Section, these functions can be found analytically in the case $\theta_0 = \pi/2$.

Comparing the first equality in Eq. (7.51) with Eq. (7.57) one can see that a correction of order $O(1)$ due to the surface energy of the particle (see Eq. (7.12)) also contributes to the free energy F_σ. As indicated in Eqs. (7.59) and (7.60), in $\gamma F_{\sigma 0}/f^2$ this correction contributes to the order $O(1)$ but in $\gamma \Delta F_\sigma / f^2$ it contributes only to the order $O(a/R_0)$ or higher. This is the case because the changes of the surface energy of the particle with respect to α occur only due to the presence of the substrate (the free energy involving a free droplet would not depend on α) which is located at a distance $O(R_0)$ from the particle and thus gives rise to the above factor $O(a/R_0)$.

7.3 Method of images for $\theta_0 = \pi/2$

For the special case $\theta_0 = \pi/2$ one can construct the Green's functions G_σ by using the method of images. First, in the reference configuration, the substrate is replaced by a virtual mirror image (with respect to the surface plane of the substrate) of the actual reference droplet. We obtain a *full* and free sphere composed of the *actual* upper hemisphere and the *virtual* lower hemisphere. Next, the deformation of the *actual* droplet subjected to a pointlike force can be constructed as the upper part of the *full* droplet subjected to this pointlike force plus additional virtual pointlike forces, subsequently called *images*, placed at the *virtual* lower hemisphere. The distribution of these images has to be chosen such that the following conditions are satisfied:

(i) the shape of the droplet, given at a direction Ω by $G_\sigma(\Omega, \Omega')$, obeys the boundary conditions in Eqs. (7.48) or (7.49), depending on the model;

(ii) the volume of the actual sessile droplet is conserved (Eq. (7.47));

(iii) the total force on the *full* droplet vanishes in the case of a pinned contact line; in the case of a free contact line the force balance is automatically satisfied by fixing the center of mass.

Condition (iii) represents a sufficient condition for mechanical equilibrium of the sessile droplet, because it implies that every piece of the full spherical interface is in equilibrium. In fact, conditions (ii) and (iii) are automatically satisfied for the Green's functions which obey Eqs. (7.44) and (7.45), because those equations have been actually obtained under the conditions of force balance and volume constraint. However, the physical assumptions as formulated in (ii) and (iii) can provide a guide for finding the distribution of images. The boundary condition expressed in (i) has to be imposed separately.

According to the above reasoning, the Green's functions for $\theta_0 = \pi/2$ can be written in the following general form:

$$G_\sigma(\Omega, \Omega') = G(\Omega, \Omega') + \sum_i Q_{\sigma i} G(\Omega, \Omega_{\sigma i}) + G_{\sigma, corr}(\Omega, \Omega'), \qquad (7.62)$$

where $G(\Omega, \Omega') = G(\Omega', \Omega)$ (see, c.f., Eq. (6.26)) is the Green's function for a *free* droplet (Eq. (6.26)). Both the amplitudes $Q_{\sigma i}$ and the directions $\Omega_{\sigma i}$ of the images, as well as a possible correction $G_{\sigma,corr}$ must be chosen such that the conditions $(i) - (iii)$ and the Green's equations (7.44) and (7.45) are satisfied.

As we shall show in the following subsections, in the case $\theta_0 = \pi/2$ the method of images can be applied successfully because the reflection at the surface plane of the substrate provides a smooth interface between the actual and virtual droplets. In the case $\theta_0 \neq \pi/2$ the interface has a non-physical cusp and the method cannot be applied directly. However, we cannot rule out the possibility that there is a modification of the method which can be used successfully also for $\theta_0 \neq \pi/2$ (for the case $\alpha = 0$ see Appendix E).

7.3.1 Free contact line

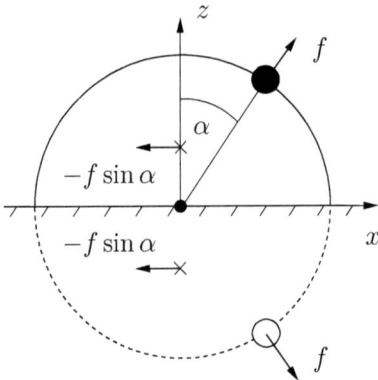

Figure 7.3: In the case of a free contact line the total drop, being the union of the actual (upper full line) and virtual (lower dashed line) drop, is pulled by f in the upper part and by its mirror image in the lower part, which corresponds to the Green's function in Eq. (7.64) (with $\Omega' = \Omega_1 = (\theta' = \alpha, \phi' = 0)$). The lateral force of magnitude $-f \sin \alpha$ fixes the center of mass ($\times$) of each hemisphere separately.

In the case of a free contact line a sessile droplet would move along the substrate if exposed to a non-vanishing lateral component of the external force applied to the particle. Therefore, in order to achieve a motion-free equilibrium, for example the lateral position of the center of mass has to be fixed. Physically, this compensation can be achieved by applying a body force to the droplet. Fixing the center of mass is to a certain extent artificial, but provides the simplest model for a free contact line.

7.3. METHOD OF IMAGES FOR $\theta_0 = \pi/2$

For $\theta_0 = \pi/2$ the boundary condition in Eq. (7.48) takes the simple form

$$\partial_\theta G_A(\Omega, \Omega')|_{\Omega \in \partial \Omega_0} = 0, \quad (7.63)$$

which can be identified with the Neumann boundary condition. Equation (7.63) and the volume constraint $\int_0^{2\pi} d\phi \int_0^{\pi/2} d\theta \sin\theta \, G_A(\Omega, \Omega') = 0$ can be satisfied by the Green's function given in Eq. (7.62) with $Q_{A1} = 1$ and $\Omega_{A1} = \hat{Z}\Omega'$ where $\hat{Z}$ means the reflection with respect to the plane $z = 0$ (see Fig. 7.3):

$$G_A(\Omega, \Omega') = G(\Omega, \Omega') + G(\Omega, \hat{Z}\Omega'), \quad (7.64)$$

which leads to

$$g_A(\alpha) = G(\bar\theta = \pi - 2\alpha). \quad (7.65)$$

with $G(\bar\theta)$ given by Eq. (6.26).

If the total drop (union of the actual and of the virtual drop) is exposed to two point forces $\pi(\Omega) = \delta(\Omega, \Omega') + \delta(\Omega, \hat{Z}\Omega')$ its resulting shape, given by Eqs. (7.46) and (7.64), renders the shape of the actual droplet for $z > 0$. For this ansatz, with $G(\Omega, \Omega')$ given by Eq. (6.26), it can be checked that $\partial_\theta G_A$ vanishes at the contact line. The correction term $G_{A,corr}$ vanishes, because the contributions to the volume stemming from the two terms in Eq. (7.64) cancel each other, so that the volume constraint is fulfilled for $G_{A,corr} \equiv 0$. As can be seen in Fig. 7.3 the total lateral force due to the pointlike forces at the interface do not vanish. Instead, the force balance is restored by fixing the center of mass (this constraint enters via the second term on the rhs of Eq. (7.44)), i.e., by applying a counterbalancing force of strength $-2\sin\theta'$ to the center of mass of the full drop in the lateral direction determined by the angle ϕ' (which means applying forces $-\sin\theta'$ both to the actual and the virtual drop, see Fig. 7.3 with $f = 1$, $\alpha = \theta'$ and $\phi' = 0$). Finally, by using Eq. (6.26) one can check explicitly that G_A as given by Eq. (7.64) satisfies Green's equation (7.44) for $\theta_0 = \pi/2$.

7.3.2 Pinned contact line

In this subsection we consider the case that the contact line is pinned at the circle corresponding to the reference configuration. One can propose that this pinning can be accomplished by substrate heterogeneities; but this requires a dedicated design in order to maintain the circular shape of the contact line. This model has the virtue that in contrast to the previous one the balance of forces is automatically satisfied without fixing the center of mass.

By choosing $Q_{B1} = -1$ positioned at $\Omega_{B1} = \hat{Z}\Omega'$ the boundary condition in Eq. (7.49) is automatically satisfied. However, the corresponding pressure field for pointlike forces placed at Ω' and at $\hat{Z}\Omega'$, which corresponds to $\pi(\Omega) = \delta(\Omega, \Omega') - \delta(\Omega, \hat{Z}\Omega')$, leads to a non-vanishing net force acting on the droplet as a whole, which is directed vertically and is equal to $2\cos\theta'$ (see Fig. 7.4). In order to restore the force balance (without fixing the center of mass) we place a second image $Q_{B2} = 2\cos\theta'$ at the bottom ("south pole", $\theta = \pi$) of the virtual droplet, so that $\Omega_{B2} = \Omega_\pi$ (see Fig. 7.4

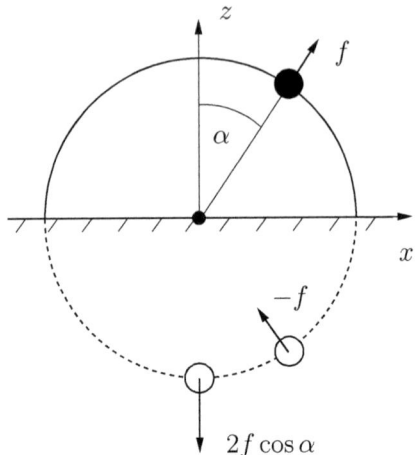

Figure 7.4: Distribution of images in the case of a pinned contact line, corresponding to the Green's function G_B in Eq. (7.67) (with $\Omega' = \Omega_1 = (\theta' = \alpha, \phi' = 0)$). The correction $G_{B,corr}$ (Eq. (7.66)) does not give rise to any forces. The center of mass is free.

with $f = 1$ and $\alpha = \theta'$). However, this second image contributes to a deformation at $\theta = \theta_0 = \pi/2$, which violates the boundary condition in Eq. (7.49). This additional contribution, which is equal to $2(\cos\theta')G(\Omega = (\theta = \pi/2, \phi), \Omega_\pi) =: -I(\theta')$ and thus depends neither on Ω nor on ϕ', has to be subtracted from the Green's function in order to uphold Eq. (7.49). Finally, the volume constraint is fulfilled by adding a second term $H(\Omega')\cos\theta$, which corresponds to a rigid vertical translation of the droplet and to an effective pressure $-(\nabla_a^2 + 2)H(\Omega')\cos\theta = -H(\Omega')(\nabla_a^2 + 2)\cos\theta$ which vanishes independently of $H(\Omega')$. Thus, this term does not contribute to the net force on the droplet. Moreover, it vanishes for $\theta = \pi/2$, i.e., at the contact line, so that it does not violate the boundary condition in Eq. (7.49). Accordingly, we are led to the following ansatz for the correction term:

$$G_{B,corr}(\Omega, \Omega') = H(\Omega')\cos\theta + I(\theta'), \tag{7.66}$$

so that the ansatz for the total Green's function is

$$G_B(\Omega, \Omega') = G(\Omega, \Omega') - G(\Omega, \hat{Z}\Omega') + 2G(\Omega, \Omega_\pi)\cos\theta' + H(\Omega')\cos\theta + I(\theta'). \tag{7.67}$$

Inserting the above expression for G_B into the boundary condition in Eq. (7.49) renders (see Appendix D)

$$I(x) = \frac{\cos x}{4\pi}. \tag{7.68}$$

The volume constraint in Eq. (7.47) provides an expression for $H(\Omega') = H(\theta')$ (see Appendix D)):

$$H(x) = \frac{1}{2\pi}\left[\cos x \ln\left(\frac{1+\cos x}{2}\right) - \cos x\right]. \tag{7.69}$$

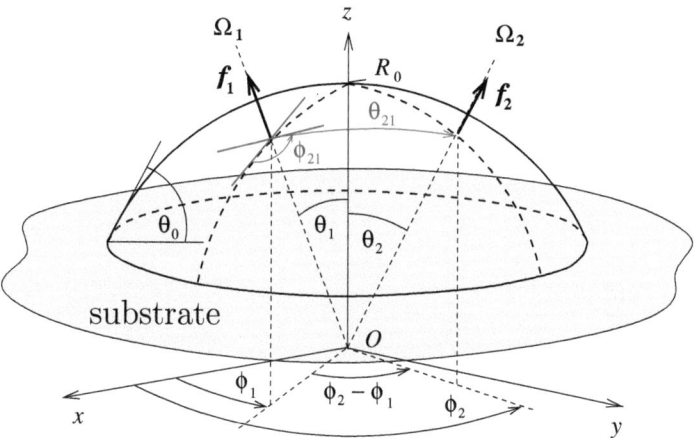

Figure 7.5: Geometry of the system and notation in the case of two particles. The dashed lines on the surface of the cap of the sphere are the points with spherical coordinates (R_0, θ_i, ϕ_i), $i = 1, 2$, with ϕ_i fixed. θ_{21} is the angle formed by the vectors $\mathbf{f}_1$ and $\mathbf{f}_2$. The angle ϕ_{21} is defined in that tangential plane to the spherical cap for which $\mathbf{f}_1$ is a normal vector. At position 1 two curves meet: one is the intersection of the plane $\phi_1 = const$ through the origin O with the spherical cap (dashed line), the other is the intersection of the spherical cap with the plane through O spanned by $\mathbf{f}_1$ and $\mathbf{f}_2$ (long red arrow); ϕ_{21} is the angle formed by the vectors tangential to these two curves.

Finally, by using Eq. (7.68), it can be checked explicitly that G_B as given by Eq. (7.67) satisfies the Green's equation (7.45) for $\theta_0 = \pi/2$; the contribution to Eq. (7.45) due to $H(\Omega')$ (see Eq. (7.67)) vanishes because $(\nabla_a^2 + 2)\cos\theta = 0$. One also obtains

$$g_B(\alpha) = -G(\bar{\theta} = \pi - 2\alpha) + 2G(\bar{\theta} = \pi - \alpha)\cos\alpha + H(\alpha)\cos\alpha + I(\alpha). \quad (7.70)$$

7.4 Many particles

It has been shown in the previous Section that the influence of a single colloidal particle on the surface of a sessile droplet can be described in terms of a capillary monopole determined by the external radial force acting on the particle. Accordingly, in the case of many particles we shall study the droplet under the action of radial external pointlike forces of magnitudes f_i giving rise to an external pressure $\Pi(\Omega) = \sum_{i=1}^{N} f_i \delta(\Omega, \Omega_i)/R_0^2$. The condition $|f_i| \ll \gamma R_0$, $i = 1, \ldots, N$ might not be sufficient for the linear theory to hold, because the forces on individual particles, even if they are small, might add up to a large total external force (and thus might lead to a large deformation); therefore we impose the more stringent condition $\sum_{i=1}^{N} |f_i| \ll \gamma R_0$. If this condition holds,

$\epsilon := \sum_{i=1}^{N} |f_i|/(\gamma R_0)$ can be regarded as a small parameter. (In the one-particle case this reduces to $\epsilon = f/(\gamma R_0)$, see Eq. (7.16)). On the other hand one has to keep in mind that the forces f_i cannot be stronger than the maximal capillary forces, which lead to the extraction of the individual particles from the surface, and which are of the order of γa_i, where a_i is the characteristic size of the particle i.

Inserting $\pi(\Omega) = \sum_{i=1}^{N} f_i \delta(\Omega, \Omega_i)/(\sum_{i=1}^{N} |f_i|)$ into Eq. (7.51) one obtains

$$F_\sigma = F_\sigma^{(N)} = \sum_{i=1}^{N} F_{i,self} + \sum_{i=1}^{N} \Delta F_{\sigma,i}^{(1)}(\theta_i, \theta_0) + \sum_{i<j} V_\sigma(\Omega_i, \Omega_j, \theta_0), \quad (7.71)$$

where on the rhs we have explicitly indicated which quantities depend on θ_0. $F_{i,self} = -f_i^2/(4\pi\gamma) \ln(R_0/a_i) + O(1)$ is the self-energy of particle i, which depends neither on the position of the particle on the droplet nor, in leading order in a_i/R_0, on the contact angle θ_0 and the boundary conditions σ. The one-particle free energy landscapes $\Delta F_{\sigma,i}^{(1)}$ are given by

$$\Delta F_{\sigma,i}^{(1)}(\theta_i, \theta_0) = -\frac{f_i^2}{2\gamma}[G_{\sigma,reg}(\Omega_i, \Omega_i, \theta_0) - G_{\sigma,reg}(0, 0, \theta_0)] = -\frac{f_i^2}{2\gamma}[g_\sigma(\theta_i, \theta_0) - g_\sigma(0, \theta_0)], \quad (7.72)$$

where the functions g_σ with $\sigma = A, B$ are independent of i and correspond to the regular part $G_{\sigma,reg}$ (i.e., not containing the logarithmic divergence) of Green's function. The last term in Eq. (7.71) consists of the effective pair potentials V_σ given by

$$V_\sigma(\Omega_i, \Omega_j, \theta_0) = -\frac{f_i f_j}{2\gamma}\Big[G_\sigma(\Omega_i, \Omega_j, \theta_0) + G_\sigma(\Omega_j, \Omega_i, \theta_0)\Big]. \quad (7.73)$$

We note that V_σ is explicitly symmetric with respect to Ω_i and Ω_j even if G_σ is not.

7.4.1 Analytical results for $\theta_0 = \pi/2$

As can be inferred from Eq. (7.72) and Fig. 7.6 the one-particle free energy landscape $\Delta F_{\sigma,i}^{(1)}(\theta_i, \theta_0 = \pi/2)$ of particle i is a non-monotonic function of its polar angle θ_i. As a consequence, besides the known phenomena of attraction of a particle to a free contact line (model A) and repulsion from a pinned one (model B), one finds a local free energy minimum for the particle being at the drop apex $\theta_i = 0$ and a local maximum at $\theta_i = \theta_{max} \approx 52°$ for model A, and a minimum at $\theta_i = \theta_{min} \approx 49°$ for model B. In the case of many particles the one-particle free energy landscapes compete with the effective pair potentials V_σ. The quantity relevant for obtaining the actual configurations of the particles is the excess free energy defined as

$$\Delta F_\sigma^{(N)} := F_\sigma^{(N)} - \sum_{i=1}^{N} F_{i,self} = \sum_{i=1}^{N} \Delta F_{\sigma,i}^{(1)} + \sum_{i<j} V_{\sigma,ij}, \quad (7.74)$$

where we have introduced the notation $V_{\sigma,ij} \equiv V_\sigma(\Omega_i, \Omega_j, \theta_0)$. In Figs. 7.7 and 7.8 we present the results for $\Delta F_\sigma^{(N)}$ in the case $N = 2$. We consider particles placed at $\Omega_1 = (\theta_1, \phi_1)$ and $\Omega_2 = (\theta_2, \phi_2)$ and subjected to external radial forces f_1 and f_2 such

7.4. MANY PARTICLES

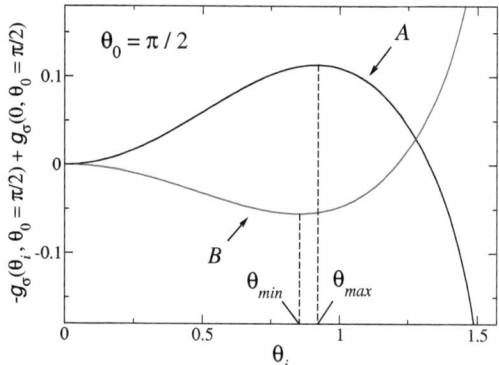

Figure 7.6: The functions $-g_\sigma(\theta_i, \theta_0 = \pi/2) + g_\sigma(0, \theta_0 = \pi/2)$, $\sigma = A, B$, which determine the behavior of the one-particle free energy landscape $\Delta F^{(1)}_{\sigma,i}(\theta_i, \theta_0 = \pi/2)$ of particle i, as functions of its polar angle θ_i, for $\theta_0 = \pi/2$ (see Eq. (7.72)).

that $|f_1| \ll \gamma R_0$, $|f_2| \ll \gamma R_0$, and $|f_1 + f_2| \ll \gamma R_0$. The excess free energy $\Delta F^{(2)}_\sigma$ is calculated for a fixed angular position of the first particle, referred to as the reference particle, as a function of the angular position of the second one acting as a probe particle. The angular cutoff δ determines the closest approach of the particles to each other, i.e., $\theta_{21} > \delta$, and to the contact line, i.e., $\theta_1, \theta_2 \in [0, \theta_0 - \delta]$, where θ_{21} is the angle between the vectors pointing in the directions Ω_1 and Ω_2 (see Fig. 7.5). One has $\delta \gtrsim 2a/R_0$, but one has to remember that for the configurations corresponding to $\theta_{21} \simeq \delta$ one should expect that the actual behavior deviates from that obtained within the monopole approximation. In the following calculations we take $\delta = \pi/36$, which corresponds to $R_0 \simeq 23a$.

The interaction potential V_σ is a function of the angular coordinates of both particles separately, i.e., it is not only a function of their separation. This is important for all angular configurations due to the long range of the capillary deformation around a monopole. This deformation does not depend on the radius R_0 of the droplet but only on the strength of the external force. In those cases in which the particles would correspond to higher capillary multipoles the pair interactions would vanish for $R_0 \to \infty$; in this sense the case of monopoles is exceptional. We also note that, as given by Eq. (7.73), V_σ is explicitly symmetric with respect to Ω_1 and Ω_2, such that the aforementioned asymmetry of Green's functions G_σ is not proliferated to the free energy.

Due to the rotational symmetry of the reference droplet forming a spherical cap the free energy depends only on the difference $\phi_2 - \phi_1$ so that $\Delta F^{(2)}_\sigma = \Delta F^{(2)}_\sigma(\theta_1, \theta_2, \phi_1 - \phi_2)$. We also introduce an auxiliary azimuthal angle ϕ_{21} which describes the angular position of the probe particle relative to the reference one (see Fig. 7.5) and thus the orientation of the pair of particles on the droplet.

In the case of a *free contact line* (Fig. 7.7) at the substrate and $\theta_1 = 0$ the minima of the free energy are degenerate both for the configurations with the probe particle at

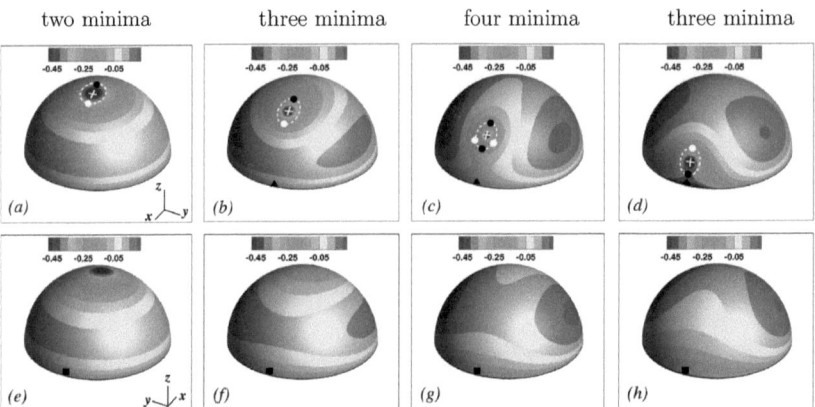

Figure 7.7: Color coded effective capillary potential for the case $\theta_0 = \pi/2$ and for a free contact line at the substrate ($\sigma = A$). The colors represent the excess surface free energy $\gamma \Delta F_A^{(2)}/(f_1 f_2)$ (Eqs. (7.71) and (7.74)) of the system as a function of the position Ω_2 of the probe particle for a fixed position Ω_1 of the reference particle (white cross). Each pair of panels in a column corresponds to the same configuration. The white dashed lines schematically indicate those positions Ω_2 of the probe particle which are at a fixed minimal polar angular separation $\theta_{21} = \delta = \pi/36$ from the reference particle, i.e., at mutual contact. Along these dashed lines $\Delta F_A^{(2)}$ is locally minimal (maximal) at the black (white) dots. The black symbols (●,■,▲) correspond to minima under the constraints $\theta_{21} > \delta$ and $\theta_2 < \pi/2 - \delta$. (a), (e)[back side]: $\Omega_1 = (\theta_1 = \pi/18, \phi_1 = 0)$; (b), (f)[back side]: $\Omega_1 = (\theta_1 = 3\pi/18, \phi_1 = 0)$; (c), (g)[back side]: $\Omega_1 = (\theta_1 = 5\pi/18, \phi_1 = 0)$; (d), (h)[back side]: $\Omega_1 = (\theta_1 = 7\pi/18, \phi_1 = 0)$. Note that in (d) two minima are shown (●,▲) separated by a potential ridge. In (c) the value $\theta_1 = 5\pi/18 = \pi/3.6$ is taken slightly smaller than $\theta_{max} \approx \pi/3.46$, which is the position of the free energy maximum for the reference particle alone (see Fig. 7.6). In such a case the upper black dot marks the deeper one of the two local minima on the dashed curve. The two local maxima (white dots) are equally high. The orientations of the coordinate axes in (b)-(d)[(f)-(h)] are the same as in (a)[(e)]. The quoted numbers of minima are the total ones occurring for the given configurations shown in the columns of panels.

7.4. MANY PARTICLES

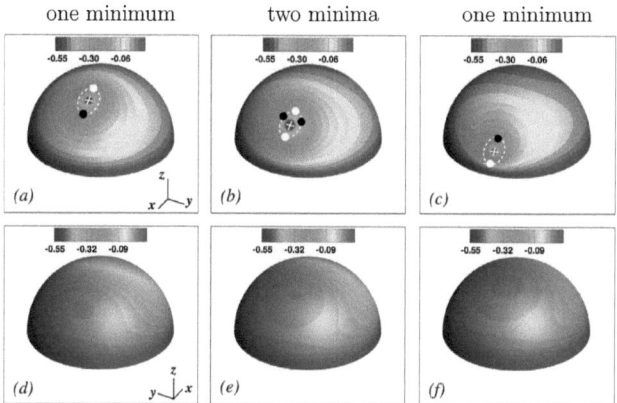

Figure 7.8: Same as Fig. 7.7 for a pinned contact line at the substrate ($\sigma = B$). (a), (d)[back side]: $\Omega_1 = (\theta_1 = 3\pi/18, \phi_1 = 0)$; (b), (e)[back side]: $\Omega_1 = (\theta_1 = 5\pi/18, \phi_1 = 0)$; (c), (f)[back side]: $\Omega_1 = (\theta_1 = 7\pi/18, \phi_1 = 0)$. In (b) the value $\theta_1 = 5\pi/18 = \pi/3.6$ is taken almost equal to $\theta_{min} \approx \pi/3.67$ which is the position of the free energy minimum for the reference particle alone (see Fig. 7.6). The positions Ω_1 (white crosses) in (a), (b), and (c) equal those in Figs. 7.7(b), (c), and (d), respectively. In (b) the two minima on the dashed white curve are degenerate. The lower white dot marks the higher local maximum on the dashed white curve.

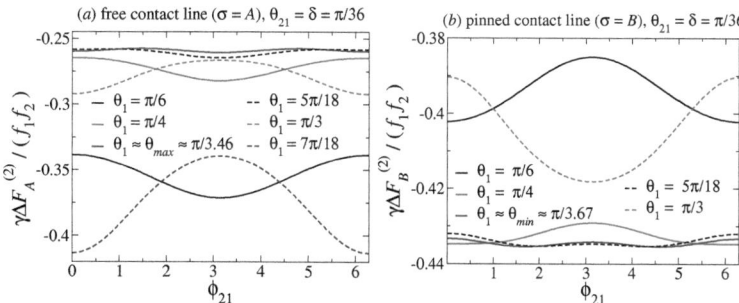

Figure 7.9: Normalized free energy $\Delta F_\sigma^{(2)}$ (Eqs. (7.71)-(7.73) and (7.74)) of a pair of colloidal particles floating on a sessile droplet with contact angle $\theta_0 = \pi/2$ and exposed to external radial forces f_i as a function of the angle ϕ_{21} (see Fig. 7.5) for several fixed values of θ_1 characterizing the polar angular position of the reference particle. Concerning θ_{max} and θ_{min} see Fig. 7.6. The polar angular separation θ_{21} (see Fig. 7.5) between two particles is kept constant upon varying ϕ_{21} and equals $\theta_{21} = \delta = \pi/36$, i.e., the free energy landscape is probed around the reference particle fixed in space. (a) free contact line; (b) pinned contact line.

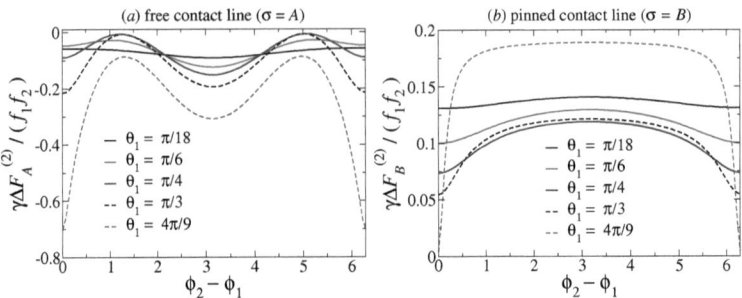

Figure 7.10: Normalized free energy (Eqs. (7.71)-(7.73) and (7.74)) of a pair of colloidal particles floating on a sessile droplet with contact angle $\theta_0 = \pi/2$ and exposed to external radial forces f_i as a function of the relative azimuthal angle $\phi_2 - \phi_1$ (see Fig. 7.5)) for several fixed polar angles θ_1 of the reference particle. Here, the free energy landscape is probed close to the substrate, i.e., $\theta_2 = \theta_0 - \delta = \pi/2 - \pi/36$. (a) free contact line; (b) pinned contact line.

$\theta_2 = \delta$ (i.e., in contact with the reference particle) and for those with $\theta_2 = \pi/2 - \delta$ (i.e., the probe particle being in contact with the substrate). In both cases the dependence on ϕ_2 is degenerate. However, if $\theta_1 \neq 0$, the rotational symmetry is broken and this degeneracy is lifted. The precise values of the angular positions of the free energy minima depend on the value of θ_1 but in general they correspond either to the closest approach of the particles to each other or to the contact line. If $\theta_1 \ll 1$, there are only two free energy minima. They occur at $\theta_{21} = \delta$, $\phi_{21} = \pi$ [●] (see $\theta_1 = \pi/6$ in Fig. 7.9(a)) and at $\theta_{21} = \pi/2 - \delta$, $\phi_2 - \phi_1 = \pi$ [■] (see $\theta_1 = \pi/18$ in Fig. 7.10(a)). Increasing θ_1 leads first to the emergence of a third minimum at $\theta_2 = \pi/2 - \delta$, $\phi_2 - \phi_1 = 0$ [▲] (see $\theta_1 = \pi/6$ in Fig. 7.10(a)) followed by the emergence of a fourth one at $\theta_{21} = \delta$, $\phi_{21} = 0$ [●] (see $\theta_1 = \theta_{max} \approx 52°$ in Fig. 7.9(a); for θ_{max} compare Fig. 7.6). Finally, upon further increasing θ_1 the minimum at $\theta_{21} = \delta$, $\phi_{21} = \pi$ turns into a local maximum, leaving three local minima at $(\theta_{21} = \delta, \phi_{21} = 0)$ [●], $(\theta_2 = \pi/2 - \delta, \phi_2 - \phi_1 = 0)$ [▲], and at $(\theta_2 = \pi/2 - \delta, \phi_2 - \phi_1 = \pi)$ [■] (see $\theta_1 = \pi/3$ in Figs. 7.9(a) and 7.10(a)). The answer to the question which configuration corresponds to the global free energy minimum with respect to the positions of *both* particles depends on δ. For $\delta \ll 1$ it is the configuration in which both particles touch the substrate and touch each other (see $\theta_1 = 4\pi/9$, $\phi_2 - \phi_1 = 0$ in Fig. 7.10(a); for $\theta_0 = \pi/2$ one has $\theta_i \leq \pi/2$), because the depth of the corresponding minimum increases as $\sim \ln(1/\delta)$. Thus the particles arrange themselves parallel to the contact line.

For a *pinned contact line* (Fig. 7.8) at the substrate we also observe a degenerate minimum of the free energy for $\theta_1 = 0$ and a broken symmetry for $\theta_1 \neq 0$. Upon further increasing θ_1, first there is only a single free energy minimum, occurring at $\theta_{21} = \delta$, $\phi_{21} = 0$ (see $\theta_1 = \pi/6$ in Fig. 7.9(b)). Upon increasing θ_1 further this minimum splits continuously into two minima at $(\theta_{21} = \delta, \phi'_{21})$ with $\phi'_{21} \in [0, \pi]$ and at $(\theta_{21} = \delta, \phi''_{21})$ with $\phi''_{21} \in [\pi, 2\pi]$ (see Fig. 7.9(b) for $\theta_1 \gtrsim \pi/4$). For $\theta_1 = \theta_{min}$ the minima have

7.4. MANY PARTICLES

reached the values $\phi'_{21} \approx \pi/2$ and $\phi''_{21} \approx 3\pi/2$, which corresponds to the configuration in which the particles are positioned parallel to the contact line (see $\theta_1 = \theta_{min} \approx 49°$ in Fig. 7.9(b)). Finally, for even larger θ_1 the two minima merge into a single minimum at $\theta_{21} = \delta$, $\phi_{21} = \pi$ (see $\theta_1 = \pi/3$ in Fig. 7.9(b)). The capillary forces repel the particles from the contact line (see Fig. 7.6) and therefore all configurations with any of the particles close to the contact line are energetically unfavorable. However, it might happen that one of the particles gets trapped close to the contact line by other means, for example due to an evaporative flux of the liquid towards the contact line or by adhesion to the substrate. If the probe particle be the trapped one, such that $\theta_2 = \pi/2 - \delta$ and ϕ_2 is free, and for any fixed polar position θ_1 of the reference particle (see Fig. 7.10(b)), the preferred position of the probe particle at the contact line always corresponds to $\phi_2 - \phi_1 = 0$, i.e., the two particles are positioned at a great circle perpendicular to the contact line. This means that for any angular position of the white cross in Fig. 7.8 the minimum of the free energy along the contact line occurs at the point closest to the white cross. In the case that both particles are constrained to lie in the neighborhood of the contact line one observes a monotonic attraction (Fig. 7.10(b), $\theta_1 = 4\pi/9$), contrary to the case of a free contact line (Fig. 7.10(a), $\theta_1 = 4\pi/9$). The configuration corresponding to the global free energy minimum with respect to the positions of *both* particles without any constraints is such that the particles touch each other at $\theta_1 = \theta_2 \approx \theta_{min} \approx 49°$ (see $\theta_1 = \theta_{min}$, $\phi_{21} = \pi \pm \pi/2$ in Fig. 7.9(b)). Thus the particles spontaneously arrange themselves parallel to the contact line at the common characteristic polar angle θ_{min}.

In summary, in both cases the particles attract each other and, as a doublet, arrange themselves such that they are both placed as close as possible to the minimum of the one-particle trapping potential $\Delta F_s^{(1)}$ which, in the case of a free contact line, occurs at the apex and at the contact line whereas in the case of a pinned contact line it occurs at an intermediate angle θ_{min}. Additionally, in the case of a free contact line there is another local free energy minimum corresponding to both particles being at the contact line, however not touching each other but being positioned on the opposite sides of the droplet ($\phi_2 - \phi_1 = \pi$).

Chapter 8

Numerical minimization of the free energy

In this Chapter we present the numerical results of the free energy minimization for particles at the surface of sessile drops. For this purpose we use the open-source numerical code Surface Evolver based on a finite element method, which incorporates surface energies as well as the volume constraint in order to find the equilibrium shapes of liquid surfaces. The numerical procedure is initiated by an arbitrarily shaped body of liquid (possibly simple, like a cube) whose boundary is subsequently divided into triangles and evolves toward the minimum energy by a gradient descent method.

In Sec. 8.1 we study the case of a spherical particle pulled (or pushed) radially by an external force f, studied analytically in Chapter 7. The condition of mechanical equilibrium, in the case of a free contact line, implies necessity of fixing the center of mass, and then, the functional to be minimized is given by Eqs. (7.4) and (7.3). In the case of a pinned contact line, the shape of the contact line is given a priori by a circle of radius $R_0 \sin\theta_0$ and thus need not be subjected to evolution. Therefore, in this case, we minimize the functional in Eq. (7.3) under the condition that those vertices of a triangulated surface, which belong to the substrate, are constrained to lie at the circle determined by the reference configuration. The numerical results fully confirm the validity of the linear theory even for droplets as small as $R_0/a = 2$.

The finite-size of the drop gives rise to an interesting effect due to the non-monotonic behavior of the Green's function $G(\bar{\theta})$ for intermediate angles. As a consequence, besides the known phenomena of attraction of the particle to the free contact line and repulsion from the pinned contact line (Kralchevsky et al. 1994), we have found a deep local free energy minimum for the particle being at the drop apex for a free contact line (Fig. 8.4) and a global free energy minimum at an intermediate angle α (Fig. 7.1) for a pinned contact line (Fig. 8.1). For forces f of the order of γa the free energy barriers associated with those minima are of the order of $0.05 \times \gamma a^2$. Thus for micron-sized particles they are typically much higher than the thermal energy, and therefore the corresponding equilibrium configurations are expected to be experimentally observable.

In Sec. 8.2 we consider the case of a free ellipsoidal particle (with $f = 0$) for $\theta_0 = \pi/2$ and various droplet sizes. The results strongly suggest that the free energy scales as $(a/R_0)^4$ which agrees with the point-quadrupole approximation (see, c.f., Eqs. (8.4) and (8.5)) studied in Chapter 6. However, in this case, the free energy changes in the regime $0 < \alpha \lesssim 45°$ are one order of magnitude smaller then in the case considered in Sec. 8.1 which leads to much bigger relative numerical errors.

8.1 Single spherical particle at a sessile droplet

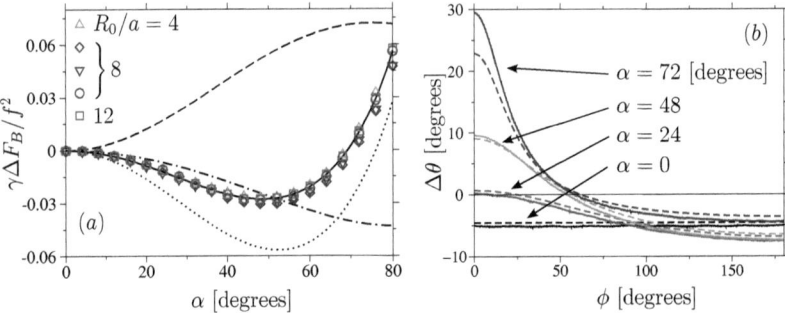

Figure 8.1: (a) Rescaled excess free energy (Eq. (7.8)) for a pinned contact line on the substrate as a function of the polar angle α (Fig. 7.1) for a contact angle $\theta_0 = \pi/2$ of the reference configuration at the substrate. The solid black curve corresponds to the analytic expression given by Eqs. (7.68)-(7.70). The dotted line and the dashed line are the contributions from the first and the second image (Fig. 7.4), respectively, and the dash-dotted line is the volume correction (Eq. (7.66)). Green triangles ($\triangle$), blue circles and red squares correspond to $\{R_a/a = 4, f/(\gamma a) = 2, \theta_p = \pi/2\}$, $\{R_0/a = 8, f/(\gamma a) = 1, \theta_p = 2\pi/3\}$, and $\{R_0/a = 12, f/(\gamma a) = 2, \theta_p = \pi/2\}$, respectively. Blue diamonds and inverted triangles ($\triangledown$) correspond to $\{R_0/a = 8, f/(\gamma a) = -1.5, \theta_p = \pi/2\}$ and $\{R_0/a = 8, f/(\gamma a) = -2, \theta_p = \pi/2\}$, respectively. (b) The difference $\Delta\theta(\phi) = \tilde{\theta}(\phi) - \theta_0$ between the actual contact angle $\tilde{\theta}(\phi)$ and $\theta_0 = \pi/2$ as a function of the azimuthal angle $\phi \in [0, \pi]$ for $\{R_a/a = 4, f/(\gamma a) = 2, \theta_p = \pi/2\}$ and for various angular positions α of the particle. The dashed lines correspond to the approximate analytic expression (Eq. (8.1)) and the solid lines are numerical results. The thin horizontal line at $\Delta\theta = 0$ corresponds to the reference configuration.

The numerical procedure follows the same protocol as the analytical approach in Chapter 7. In order to determine the dependence of the excess free energy ΔF in Eq. (7.8) on α, we fix the angular position α of the particle and minimize the free energy functional given in Eq. (7.3) with respect to the radial displacement of the particle h and the shape of the droplet under the constrains of a fixed center of mass (Eq. (7.4)) or a pinned contact line (Eq. (7.7)). In order to obtain the configurations with $\alpha \neq 0$ the surface is pre-evolved for $\alpha = 0$ from an initial shape and then the particle is moved

8.1. SINGLE SPHERICAL PARTICLE

in a step-wise fashion by small increments $\Delta\alpha$. In both models A and B the contact line at the particle is taken to be free. In the case of model B the contact line at the substrate is pinned from the very beginning at a circle corresponding to the reference configuration with $f = 0$. The geometrical center of the reference droplet, from which the radial distance of the particle is measured, is well defined throughout the evolution. In the case of model A, the situation is more complicated because, due to the finite size of the particle, the position $x_{CM,ref}$ of the center of mass of liquid depends on α. This finite-size effect must be taken into account for medium-sized droplets (see, c.f., Eqs. (8.2) and (8.3)).

8.1.1 Pinned contact line on the substrate

In Fig. 8.1 we plot numerical values of $\gamma\Delta F_B/f^2$ for various droplet sizes R_0, both signs and various strengths of f, and various contact angles θ_p at the particle. The variation of all these parameters does not affect the α-dependence and we obtain a single master curve, in very good agreement with the theoretical expression in Eq. (7.72) for a pointlike force. This master curve exhibits a global minimum at the angle $\alpha_{min} \approx 48°$, which corresponds to the global equilibrium configuration, independent of f, θ_p and R_0 (see Fig. 8.3(a) and (b)). Finally, we note that from Fig. 8.1 it also follows that, in the theoretical expression for the Green's function G_B (Eq. (7.67)) the contributions from both images and from the volume correction $G_{B,corr}$ are all equally important.

We emphasize the de facto independence of ΔF from the contact angle θ_p at the particle, which justifies the monopole approximation in which we neglect the changes of the wetting energy at the particle. The only systematic deviation from the analytic expression occurs for negative values of f. We have investigated the system for various absolute values of negative f and the free energy turns out to be always slightly overestimated by the analytic theory. In fact we obtain two master curves, which are close to each other: one for positive and one for negative f.

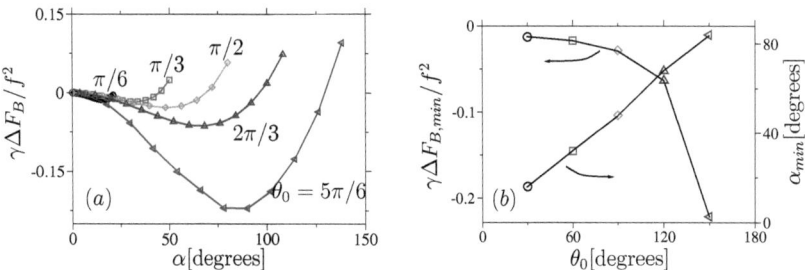

Figure 8.2: (a) Dependence of the rescaled excess free energy (Eq. (7.8)) on the contact angle θ_0 and a pinned contact line on the substrate. For all curves $f/(\gamma a) = 2, \theta_p = \pi/2$, and $V_l = 79 \times \frac{4\pi}{3}a^3$. (b) Position and depth of the free energy minimum as functions of θ_0.

114 CHAPTER 8. NUMERICAL MINIMIZATION OF THE FREE ENERGY

For $f \neq 0$ the actual contact angle $\tilde{\theta}(\phi)$ at a pinned contact line (see Eq. (7.31)) differs from the constant value θ_0 for the reference configuration ($f = 0$). Up to first order in ϵ one obtains for $\theta_0 = \pi/2$ and with $v(\Omega) = q\, G_B(\Omega, \Omega_1)$

$$\cos\tilde{\theta}(\phi) = \frac{a}{R_0} \frac{f}{\gamma a} \partial_\theta G_B(\Omega, \Omega_1)|_{\Omega=(\theta=\pi/2,\phi)}. \tag{8.1}$$

The numerical results for $\Delta\theta(\phi) := \tilde{\theta}(\phi) - \pi/2$ and the comparison with Eq. (8.1) are presented in Fig. 8.1(b). The discrepancy is typically of the order of 10%, which differs from the almost perfect agreement found in the case of the free energy (see Fig. 8.1(a)). If the particle is close to the contact line ($\alpha = 72°$) the discrepancy reaches 25% which is linked to the large value of $\Delta\tilde{\theta}(\phi) \approx 30° \approx 0.5$ [rad] for $\phi \approx 0$, which in turn signals that in the vicinity of the contact line and close to the particle the small gradient approximation breaks down and the terms $O(\epsilon^2)$ in Eq. (7.31) become important (for the values of the parameters used in Fig. 8.1(b) one has $\epsilon = 0.5$).

On the other hand we observe a strong dependence on θ_0. For the contact angle on the substrate being different from $\theta_0 = \pi/2$ the minimum is always in between the apex and the contact line on the substrate (see Fig. 8.2(a)), just like in the case $\theta - 0 = \pi/2$, whereas its depth increases strongly with θ_0 approaching π (Fig. 8.2(b)). The droplet shapes corresponding to the global free energy minima for various values of θ_0 have been displayed in Fig. 8.3.

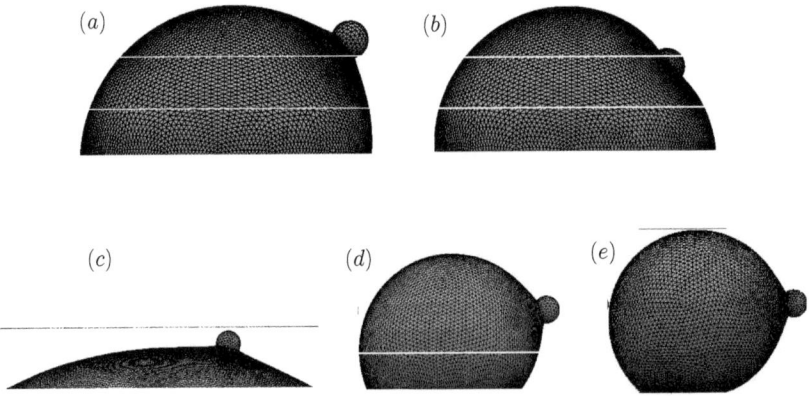

Figure 8.3: Droplet shapes corresponding to the global free energy minima for various droplet geometries for a pinned contact line at the substrate, and a given volume of liquid (the same for all configurations). $(a), (b)$: $\theta_p = \pi/2$; the minimum of $\Delta F(\alpha)$ occurs at $\alpha \simeq 48°$ independently of the pulling force (here $f_0/(\gamma a) = 2$ in (a), and $f_0/(\gamma a) = -2$ in (b)) and droplet radius (here $R_0/a = 8$). In the remaining configurations $f_0/(\gamma a) = 2$ and $\theta_p = \pi/6$ (c), $\theta_p = 2\pi/3$ (d) and $\theta_p = 5\pi/6$ (e). Note a different geometric contact angle at the substrate on different sides of the droplet.

8.1.2 Free contact line on the substrate

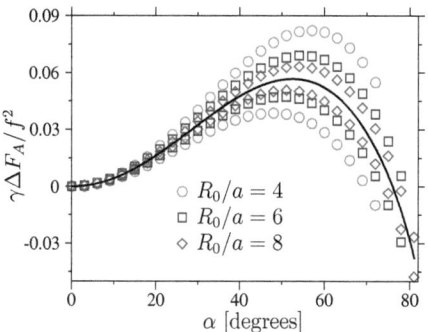

Figure 8.4: The rescaled excess free energy (Eq. (7.8)) as a function of the polar angle α for the contact angle $\theta_0 = \pi/2$ of a free contact line at the substrate, $f/(\gamma a) = 1$ (below the solid line) and $f/(\gamma a) = -1$ (above the solid line), and various radii R_0. The solid line corresponds to the expression given by Eqs. (6.26), (7.72), and (7.65) for pointlike forces. For geometrical reasons the maximal accessible values of α decrease for decreasing R_0. For a free contact line the contact angle at the substrate remains at the value θ_0 for $f \neq 0$, i.e., it remains the same as for the reference configuration.

The results for a free contact line, presented in Fig. 8.4, demonstrate that changing the boundary conditions on the substrate can change the behavior of the particle completely. The equilibrium position is now at the contact line and there is a deep metastable free energy minimum at the drop apex (see Fig. 8.5).

In the case of a free contact line we observe much larger discrepancies with the predictions of the theory for pointlike forces, revealing a dependence of ΔF_A on f beyond the simple scaling $\sim f^2$. However, this deviation vanishes for increasing radii R_0 of the droplet indicating that this is a finite-size effect. In order to understand this effect we first consider the reference configuration, i.e., the case $f = 0$. For $\theta_p = \pi/2$ the immersed part ΔV of the particle (being the intersection of the domain occupied by the particle with the spherical cap representing the reference droplet of volume approximately $V_l = 2\pi R_0^3/3$ in the case $\theta_0 = \pi/2$) for $\theta_p = \pi/2$ has the constant volume of approximately $\Delta V = 2\pi a^3/3$, but its position depends on α. Therefore $x_{CM,ref}$ also depends on α and equals (here the position of the particle is taken to have a positive x component, see Fig. 7.1):

$$x_{CM,ref}(\alpha) = \frac{\int_{V_l} dV\, x}{V_l} = -\frac{1}{V_l}\int_{\Delta V} dV\, x \approx -\frac{\Delta V}{V_l} R_0 \sin\alpha \approx -\left(\frac{a}{R_0}\right)^2 a \sin\alpha. \quad (8.2)$$

In contrast, in the case of a pointlike force one has $x_{CM,ref} \equiv 0$. The corresponding difference in the free energy δF can be understood as the total work done by the force $f_{CM}(\alpha) = -f \sin\alpha$ (see the main text after Eq. (7.40)) applied to the center of mass in order to counterbalance the lateral component of the force f (see Eq. (7.40)), upon

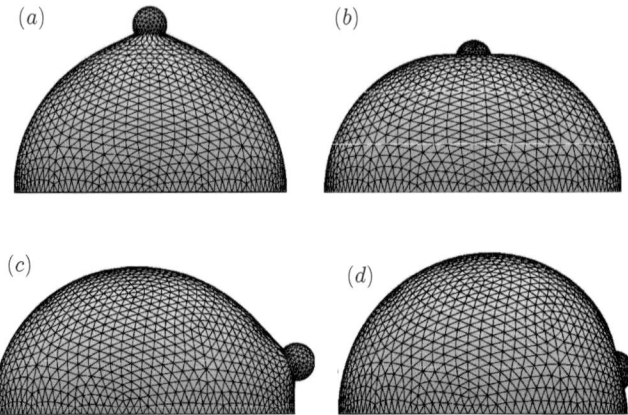

Figure 8.5: (a) and (b): droplet shapes corresponding to the local free energy minima at the apex for a free contact line at the substrate and $\theta_0 = \pi/2$; (c) and (d): configurations with the particle close to the global minima (at the contact line); $f_0/(\gamma a) = 2$ in (a), (c), and $f_0/(\gamma a) = -2$ in (b), (d). Note the constant contact angle at the substrate.

displacing the center of mass by $-x_{CM,ref}(\alpha)$:

$$\delta F = \int_0^\alpha d\alpha' \, f_{CM}(\alpha')(-dx_{CM,ref}/d\alpha') =$$
$$= -fa\left(\frac{a}{R_0}\right)^2 \int_0^\alpha d\alpha' \sin\alpha' \cos\alpha' = -\frac{fa}{2}\left(\frac{a}{R_0}\right)^2 \sin^2\alpha. \quad (8.3)$$

This correction should be subtracted from the numerically calculated free energy in order to facilitate the comparison with the analytical result for a pointlike force (we have ignored the correction δx, which also depends on α, see Eq. (7.13), but gives a contribution of the order $(f^2/\gamma) \times O(a/R_0)^3)$. It is linear in f, which explains the aforementioned deviation from the scaling $\sim f^2$. For large drops it vanishes $\sim (a/R_0)^2$ so that for $(a/R_0) \lesssim 0.1$ it can practically be neglected (see Fig. 8.4). However, for smaller droplets this correction has to be taken into account in order to obtain agreement with the perturbation theory (see Fig. 8.6(a)).

8.2 Single ellipsoidal particle at a sessile droplet

The results of the previous Section have shown that the free energy landscape ΔF of a single spherical particle subjected to an external radial force is well reproduced by a point-force approximation. In this Section we use the analytical results for the interaction energy of pointlike quadrupoles derived in Chapter 6 in order to compare with the numerical calculations of the free energy landscape ΔF_{el} of a free ellipsoidal particle ($f = 0$) at a sessile half-spherical droplet, see Fig. 8.7. In analogy to the case

8.2. SINGLE ELLIPSOIDAL PARTICLE

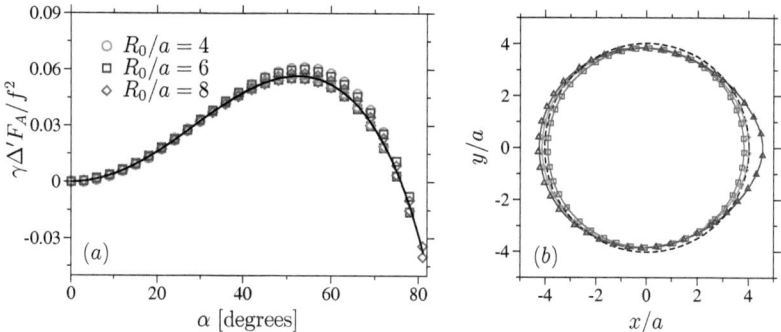

Figure 8.6: (a) The rescaled excess free energy (Eq. (7.8)) for a free contact line at the substrate with taking into account the finite size correction δF (Eq. (8.3)) for the numerical data (symbols): $\Delta' F = \Delta F - \delta F$; the color code is the same as in Fig. 8.4, $f/(\gamma a) = 1$ and $f/(\gamma a) = -1$ for the points slightly below and above the solid line, respectively (valid for $\alpha > 60°$). (b) The deformation of a free contact line at the substrate for $\alpha = 12°$ (squares), $\alpha = 48°$ (diamonds), and $\alpha = 72°$ (triangles), where $\theta_0 = \pi/2, R_0/a = 4, f/(\gamma a) = 2$, and $\theta_p = \pi/2$. The dashed line denotes the reference configuration. The solid lines are predicted analytically for pointlike forces by Eqs. (7.46), (6.26), and (7.64).

of spherical particles, we expect that the point-multipole approximation should be valid as long as the deformation of the interface outside the ellipsoidal particle can be reproduced by a constant pointlike quadrupole Q_2 placed in the center of the particle. For the configurations with the particle very close to the contact line one would expect that Q_2 is no longer constant and moreover that the higher multipoles may become important.

We performed the numerical calculations for $\theta_0 = \pi/2$ for a free and a pinned contact line at the substrate, see Fig. 8.8. In the case of a free contact line the substrate acts as a "mirror" and the shape of a sessile droplet is the same as the shape of a half of a full droplet with two identical particles in a configuration reflecting the mirror symmetry. Thus method of images in this case can be directly applied. Approximating the particle as a point quadrupole Q_2, oriented with the angle ϕ_1 with respect to the great circle passing through it and perpendicular to the substrate, the image is also a point quadrupole Q_2 oriented such that $\phi_2 = \pi - \phi_1$. The energy of a particle at a sessile droplet is a half of the energy of two particles at a full droplet. Thus, according to Eq. (6.54) and taking $\theta_{12} = \pi - 2\alpha$, we obtain an approximate expression for the excess free energy $\Delta F_{el,A}$ for a free contact line at the substrate in the form

$$\Delta F_{el,A}(\alpha) = -\frac{3\pi}{32}\gamma(\Delta u_{max})^2 \left(\frac{a}{R_0}\right)^4 (\cos^{-4}\alpha - 1). \tag{8.4}$$

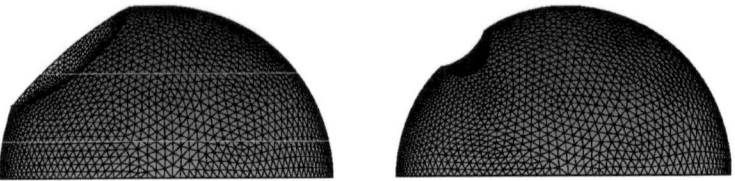

Figure 8.7: Sections of the droplet shapes for a prolate ellipsoid with contact angle $\theta_p = \pi/2$ placed at the droplet surface in a configuration with the particle main axis directed perpendicular (left panel) or parallel (right panel) to the contact line at the substrate; droplet radius is $R_0/a = 6$ and the particle main axes are $(3a, a, a)$; contact line at the substrate is pinned.

which yields a monotonic attraction of the particle towards a contact line (see Fig. 8.8(a) and 8.9). Note, that the free energy in the point-quadrupole approximation does not depend on the orientation of the particle ϕ, because $\cos(2\phi_1 + 2\phi_2) = \cos 2\pi = 1$. This is in general not true for an actual ellipsoidal particle due to non-vanishing contributions from higher multipoles (or actually from the elliptic multipoles Lehle 2007), but here we do not investigate the dependence of the free energy on ϕ.

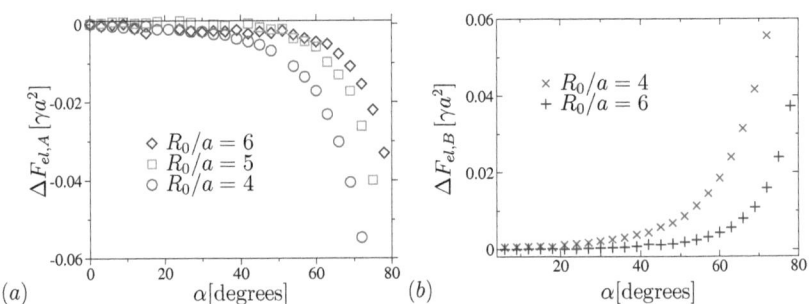

Figure 8.8: Free energy of a prolate ellipsoid with main axes $(3a, a, a)$ and the contact angle $\theta_p = \pi/2$ at the surface of a sessile droplet of various radii, with a free (a) or a pinned (b) contact line at the substrate and $\theta_0 = \pi/2$.

In the case of a pinned contact line at the substrate Green's function G_B contains a contribution from an additional image and a correction term $G_{B,corr}$ (Eq. (7.66)) responsible for the volume constraint. In the simplest estimate we neglect this correction and approximate the excess free energy $\Delta F_{el,B}$ only by the interaction energy with a single image quadrupole $-Q_2$ (oriented the same as in model A), such that

$$\Delta F_{el,B}(\alpha) \approx -\Delta F_{el,A}(\alpha) = \frac{3\pi}{32}\gamma(\Delta u_{max})^2 \left(\frac{a}{R_0}\right)^4 (\cos^{-4}\alpha - 1), \qquad (8.5)$$

8.2. SINGLE ELLIPSOIDAL PARTICLE

which yields a monotonic repulsion from the contact line, see Figs. 8.8(b) and 8.9. In both cases of a free and a pinned contact line we have not observed any local free energy minima, as we have in the case for a capillary monopole. The reason for this can be understood by considering again the deformations of a free droplet. In the presence of a capillary monopole, the droplet is stretched at the poles and simultaneously squeezed at the equator in order to maintain the constant volume. This non-monotonicity of the deformation results in a non-monotonic interaction free energy with another particle. In the case of a quadrupole the deformation of the droplet is not axisymmetric. The contributions to the volume $\sim \cos 2\phi$ coming from different directions ϕ cancel each other (that is actually the case for all multipoles of order higher that $m = 0$) and the deformation of the droplet is monotonically decaying with θ, which as a consequence leads to a monotonic interaction free energy.

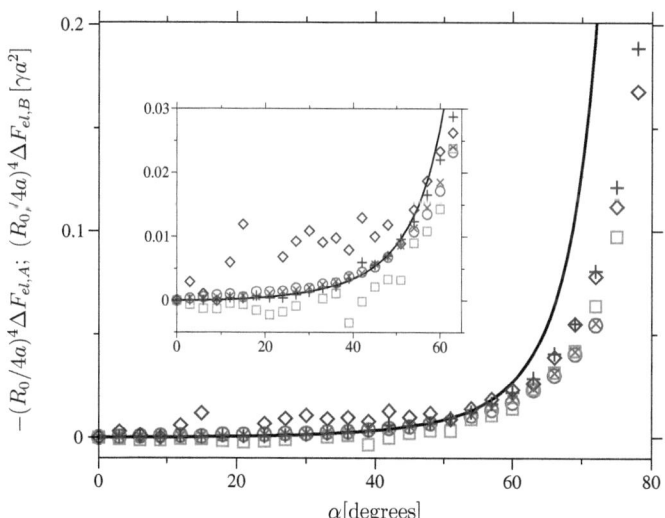

Figure 8.9: $-\Delta F_{el,A}$ and $\Delta F_{el,B}$ (Fig. 8.8) rescaled with $(R_0/4a)^4$. Note the collapse of the data onto the theoretical expression in Eq. 8.4 (with $\Delta u_{max}/a = 1.3$ fitted to the data for $R_0/a = 4$) for $\alpha \lesssim 60°$. The data for $R_0/a = 5$ and $R_0/a = 6$ in the case of model A (squares and diamonds, respectively) for $\alpha \lesssim 50°$ suffer from large numerical errors.

In Fig. 8.8(a) we present the results for an ellipsoidal particle with half-axes $(3a, a, a)$ oriented parallel to the contact line at the substrate and for droplets of radii $R_0/a = 4, 5$ and 6 with a free contact line at the substrate. The contact angle at the particle is taken $\theta_p = \pi/2$, which means that in the case of a flat liquid surface the particle would not induce any deformation of the interface ($\Delta u_{max} \equiv 0$). However, in the case of a

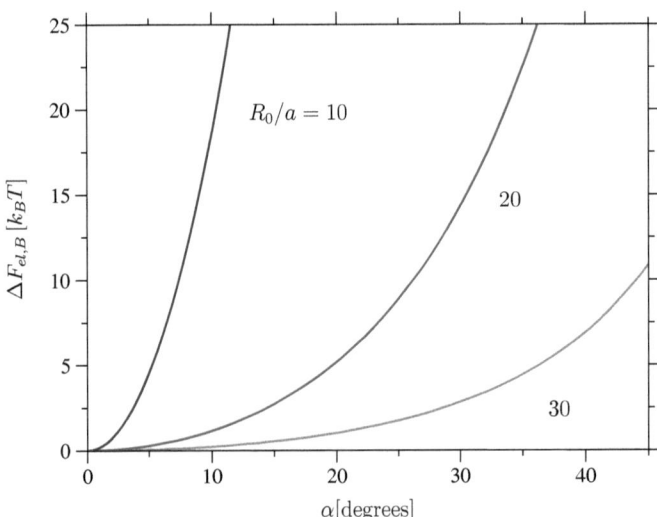

Figure 8.10: Confining potential (Eq. (8.5)) for an ellipsoidal particle of size $a = 1\mu m$ at the surface of a half-spherical sessile droplet with a pinned contact line for various values of R_0/a. The undulation of the contact line is assumed constant and equal $\Delta u_{max}/a = 1$.

spherical interface of radius R_0 the particle deforms the interface and we can expect that the maximal deformation Δu_{max} must depend on R_0, such that it vanishes in the limit of a flat interface, i.e., for $R_0 \to \infty$. Thus, for very large droplets, one could assume a linear relation in which the deformation Δu_{max}, as compared with the particle size a, is proportional to the ratio a/R_0 which would give $\Delta u_{max} \sim a^2/R_0$. Thus, for $\theta_p = \pi/2$, we would expect the excess free energy to scale as $(a/R_0)^6$. However, this does not necessarily hold for intermediate-sized droplets and one can assume that scaling of the actual free energy is intermediate between $(a/R_0)^6$ (for $\Delta u_{max} \sim a^2/R_0$) and $(a/R_0)^4$ (for $\Delta u_{max} = const$). The numerical data indicate that in the considered droplet size regime the scaling $(a/R_0)^4$ is more likely reproduced, see Fig. 8.9. The fitted value of $\Delta u_{max}/a$ is 1.3 which seems large, but one has to remember that this undulation should be measured to a fixed distance $r = a$ from the particle center. Therefore, the actual undulation of the contact line at the ellipsoidal particle of main axes $(3a, a, a)$ is much smaller (the numerically calculated undulation of the contact line equals approximately $\Delta u_{max,num}/a \approx 0.5$).

In the case of a pinned contact line we observe a collapse of the data $\Delta F_{el,B}$ with $-\Delta F_{el,A}$ for a given droplet radius (see Fig. 8.9). This indicates that any possible

8.2. SINGLE ELLIPSOIDAL PARTICLE

corrections (for example due to the volume constraint) to Eq. 8.5 must be small. The data again suggest a possible scaling $(a/R_0)^4$.

In agreement with a point-quadrupole model we observe the repulsion from a pinned contact line corresponding to the energy barrier for $R_0/a = 4$ and $\alpha < \pi/4$ of the order $10^{-5}\gamma a^2 = 10^2 k_B T \times (a/1\mu m)^2$, such that the effect of entrapment of the particle at the drop apex could be of practical relevance for micro-particles. However, the numerical value of the barrier is very sensitive to the droplet radius (as it scales as $(a/R_0)^4$). In order to indicate this dependence in Fig. 8.10 we plot the confining potentials $\Delta F_{el,B}$ (Eq. (8.5)) in units of $k_B T$ for $\gamma = 0.05 N/m$ and various values of the ratio R_0/a. We assume the contact angle at the particle to be different that $\pi/2$ such that the undulation of the contact line measured by Δu_{max} remains constant with $R_0 \to \infty$. With growing R_0 the energy landscape significantly flattens but still it can lead to entrapment (with a barrier of $10 k_B T$) of non-spherical particles of size $a = 1\mu m$ in the region $\alpha < \pi/4$ even for droplets as large as $R_0/a = 30$ (or $R_0/a = 100$ for $a = 10\mu m$).

Chapter 9

Summary

The results of the thesis may be summarized as follows.

We have studied the behavior of particles at liquid-gas interfaces in the situation when the particles are partially wetted by the liquid. First, we have investigated the case of a flat interface. We have derived exact expressions (Eqs. (3.11)-(3.13)) for the free energy depending on the immersion of the particle in the liquid phase. We have found that the exact results are almost perfectly reproduced by a linearized theory for small deformations of the flat interface with the amplitude renormalized in order to match the solution of the full non-linear problem far away from the particle. Furthermore, we have compared the above macroscopic approach with a microscopic calculations based on the mean-field density functional theory for the fluid surrounding the particle being characterized by long-range intermolecular forces and assuming so-called sharp-kink of the density profile at the interface. From these results we have drawn a general conclusion that the macroscopic linear theory is sufficient for the purpose of calculating the capillary forces even for sub-micrometer particles and that the effect of the long-range intermolecular forces enters then only through the surface tensions.

Next, we have investigated the interaction free energy of two heavy spheres at a flat fluid-fluid interface in presence of gravity. We have assumed that the contact lines at the particles are pinned which leads to the asymptotic result $f^2 \ln(qd)/(2\pi\gamma)$ as a function of the spatial separation d of the particles, where f is the effective weight of the particles and q is the inverse capillary length, which coincides with the result obtained by Oettel et al. (2005b) for the particles with free contact lines (fixed contact angles). This leads to a conclusion that, for the leading asymptotic behavior, the mechanism of the attachment of the particles to the interface is irrelevant.

In order to set up a more general theoretical framework, we have introduced the analogy between capillarity and electrostatics, developed recently by Domínguez et al. (2008a), in which the small deformations of an initially flat interface play the role of the electrostatic potential and the external pressure can be interpreted as the "capillary charge" distribution. In the case of a deformation induced by a particle the analogy can be used to identify the capillary monopole with the total external force acting on the particle and the capillary dipole with the total external torque. As a consequence, a free particle of arbitrary shape corresponds to a quadrupole. In this picture the asymptotic results for the interaction energy between particles subjected to external forces or between free ellipsoidal particles, reported in the literature, can be easily explained in terms of electrostatics and multipole expansion.

Subsequently, we have studied the spherical interfaces. We have considered small deformations of a spherical droplet subjected to an external pressure field Π. In the limit of large droplets we have obtained a relation between the spherical multipoles Π_{lm} associated with a particle and the capillary multipoles $Q_{|m|}$ for the identical particle at a flat interface. We have derived general expressions for the interaction potentials between arbitrary-order multipoles at an arbitrary angular separation. We have shown that the result for monopoles reproduces the Green's function derived by Morse & Witten (1993). Additionally, we have obtained a closed expression for point-quadrupoles (see Eq. (6.51) and Appendix D)).

Finally, we have approached the problem of a single spherical particle at the surface of a *sessile* droplet. In the cases without axial symmetry we have taken into account the fact that the condition of balance of forces acting on the droplet in the lateral direction requires either a fixed lateral position of the center of mass of the droplet (model A) or a pinned contact line at the substrate (model B). Using a perturbation theory for small deformations of the reference cap-like spherical shape of the droplet we have derived the free energy functional incorporating the liquid-substrate surface free energy (see Appendix B). The effects of the particle being pulled (or pushed) by an external force f and of the fixed center of mass have been incorporated by introducing effective pressure fields π and π_{CM}, respectively. In terms of those fields the linear Young-Laplace equation governing the small deformations has been derived (Eq. (7.26)). We have shown that in the limit of a small particle the free energy of the sessile droplet (Eq. (7.51)) can be expressed in terms of the Green's functions satisfying either the Robin or the Dirichlet boundary conditions at the substrate corresponding to a free or a pinned contact line (Eqs. (7.48) and (7.49)), respectively. The free energy does not depend on the size of the particle but only on the pulling force f (Eq. (7.58)), the contact angle θ_0 at the substrate, and on the angular position of the particle α. For $\theta_0 = \pi/2$ we have exploited an analogue of the method of images known from electrostatics in order to calculate the surface free energy (in excess over the surface free energy of the reference configuration) as a function of α analytically. Because in this case the reference droplet forms a half of a sphere the boundary conditions at the substrate can be fulfilled by introducing an image particle at the virtual hemisphere below the substrate surface (such that the union of the actual and the virtual droplet forms a full sphere). Further analysis shows that due to the conditions of force balance and volume constraint the Green's function requires additional terms (see Eq. (7.67)), but they do not change the results qualitatively (see various contributions in Fig. 8.1(b)). Using the analytical results for the Green's functions (see Sec. 7.3 and Appendix D) in the case $\theta_0 = \pi/2$ we have also calculated pair-potentials for two particles at arbitrary angular positions at the droplet and analyzed possible minimum free energy configurations.

The analytical results have been compared with the results of the numerical minimization of the free energy functionals for a spherical and for an ellipsoidal particle at a sessile droplet. In the case of a spherical particle subjected to an external force we have found an almost perfect agreement with the predictions of the perturbation theory in the case $\theta_0 = \pi/2$. Besides the known phenomena of attraction of a particle to a free contact line and repulsion from a pinned one, we have observed a local free energy minimum for the particle being located at the drop apex (Fig. 8.4) or at a characteristic intermediate angle (Fig. 8.1(a)), respectively. This peculiarity can be traced back to a non-monotonic behavior of the Green's functions for a free droplet, which is a consequence of interplay between the deformations of the droplet and the volume constraint.

In the case of force-free ellipsoidal particles we have obtained monotonic free energy landscapes, in qualitative and partially quantitative agreement with the point-quadrupole approximation (Fig. 8.9). Particularly, the theoretically predicted monotonic dependence of the free energy on α and scaling $\sim (a/R_0)^4$ has been confirmed. We have argued that in the case of a pinned contact line at the substrate the equilibrium configuration with the particle at the drop apex is robust against thermal fluctuations and therefore it could be observed in an experiment (Fig. 8.10).

We note that the observed dependence on θ_0 (Fig. 8.2) for particles subjected to an external force remains an open problem (even qualitative). As a further outlook, the pair potential ΔF_{22} for point-quadrupoles at a free droplet could be used in order to derive the corresponding pair-potential in the case of a sessile droplet, which could be of significant practical importance, because, as we have shown, force- and torque-free particles trapped at the surface of a drop correspond to capillary quadrupoles. In much more general terms, it is also still a matter of a future research to extend the theory of capillary interactions beyond flat and spherical interfaces towards general curved interfaces, which could find its applications in mixtures of immiscible liquids and in phase separating fluids.

Appendix A

Calculation of the interaction free energy ΔF_{22}

The Wigner d-matrix can be represented in terms of the Jacobi polynomials $P_n^{(\alpha,\beta)}$ (see, for example Edmonds 1957):

$$d^l_{m',m}(\beta) = \left[\frac{(l+m)!(l-m)!}{(l+m')!(l-m')!}\right]^{1/2} \left(\sin\frac{\beta}{2}\right)^{m-m'} \left(\cos\frac{\beta}{2}\right)^{m+m'} P_{l-m}^{(m-m',m+m')}(\cos\beta) \tag{A.1}$$

and the needed elements of the Wigner d-matrix equal

$$d^l_{-2,2}(\theta) = \left(\frac{1-\cos\theta}{2}\right)^2 P_{l-2}^{(4,0)}(\cos\theta), \tag{A.2}$$

$$d^l_{2,2}(\theta) = \left(\frac{1+\cos\theta}{2}\right)^2 P_{l-2}^{(0,4)}(\cos\theta), \tag{A.3}$$

$$\tag{A.4}$$

Inserting the above expressions into Eq. (6.50) we obtain

$$\Delta F_{22}(\theta,\phi_1,\phi_2) = -\gamma a^2 \left(\frac{a}{R_0}\right)^4 \frac{Q_{1,2}}{\gamma a^3} \frac{Q_{2,2}}{\gamma a^3} \frac{1}{128\pi}$$

$$\times \left[\cos(2\phi_1+2\phi_2)\left(\frac{1-\cos\theta_{12}}{2}\right)^2 \sum_{n=0}^{\infty}(2n+5)\frac{(n+3)!}{(n+1)!}P_n^{(4,0)}(\cos\theta_{12})\right.$$

$$\left. +\cos(2\phi_1-2\phi_2)\left(\frac{1+\cos\theta_{12}}{2}\right)^2 \sum_{n=0}^{\infty}(2n+5)\frac{(n+3)!}{(n+1)!}P_n^{(0,4)}(\cos\theta_{12})\right]. \tag{A.5}$$

The series entering the above expression can be evaluated using the generating function $g(x,z)$ for the Jacobi polynomials, which reads (Abramowitz & Stegun 1970):

$$g(x,z) = \frac{1}{R(1-z+R)^\alpha(1+z+R)^\beta} = 2^{-\alpha-\beta}\sum_{n=0}^{\infty}z^n P_n^{(\alpha,\beta)}(x), \quad |z|<1 \tag{A.6}$$

where
$$R = \sqrt{1 - 2xz + z^2}. \tag{A.7}$$

We introduce the following auxiliary functions expressed in terms of the derivatives of the generating function $g(x, z)$

$$g_0(x) = \sum_{n=0} P_n^{(\alpha,\beta)}(x) = 2^{\alpha+\beta} \lim_{z \to 1} g(x, z), \tag{A.8}$$

$$g_1(x) = \sum_{n=0} n P_n^{(\alpha,\beta)}(x) = 2^{\alpha+\beta} \lim_{z \to 1} \frac{\partial}{\partial z} g(x, z), \tag{A.9}$$

$$g_2(x) = \sum_{n=0} n^2 P_n^{(\alpha,\beta)}(x) = 2^{\alpha+\beta} \lim_{z \to 1} \frac{\partial^2}{\partial z^2} g(x, z) + g_1(x), \tag{A.10}$$

$$g_3(x) = \sum_{n=0} n^3 P_n^{(\alpha,\beta)}(x) = 2^{\alpha+\beta} \lim_{z \to 1} \frac{\partial^3}{\partial z^3} g(x, z) + 3g_2(x) - 2g_1(x). \tag{A.11}$$

The series appearing in Eq. (A.5) can be then simplified as

$$\sum_{n=0}(2n+5)\frac{(n+3)!}{(n+1)!}P_n^{(\alpha,\beta)}(x) = \sum_{n=0}(2n^3 + 15n^2 + 37n + 30)P_n^{(\alpha,\beta)}(x)$$
$$= 2^{\alpha+\beta}[2g_3(x) + 15g_2(x) + 37g_1(x) + 30g_0(x)], \tag{A.12}$$

After some algebra we obtain

$$\sum_{n=0}(2n+5)(n+3)(n+2)P_n^{(4,0)}(x) = \frac{96}{(1-x)^4}, \tag{A.13}$$

$$\sum_{n=0}(2n+5)(n+3)(n+2)P_n^{(0,4)}(x) = 0, \tag{A.14}$$

which together with Eq. (A.5) finally yields Eq. (6.51).

Appendix B

Derivation of the free energy functional $\mathcal{F}[\{v(\Omega)\}]$ for a sessile droplet

Our starting point is the exact functional in Eq. (7.9). In order to separate the boundary terms we would like to split $\mathcal{F}$ into two parts: one composed of integrals over Ω_0 and the other one incorporating integrals over $\Omega_c \setminus \Omega_0$ and $\Omega_0 \setminus \Omega_c$. However, the deformation $u(\Omega)$ is not defined inside the domain $\Omega_0 \setminus \Omega_c$. Therefore it is more convenient first to apply the ϵ-expansion in terms of which the dimensionless deformation $v(\Omega)$ can be linearly extrapolated into $\Omega_0 \setminus \Omega_c$ and then to proceed with the decomposition of the free energy functional into the surface and the boundary terms.

First, we note that $\theta_c(\phi) - \theta_0 = O(\epsilon)$ and thus, because we want to keep only the terms up to second order in ϵ in the free energy functional, we can neglect the terms $O(\epsilon^2)$ under the integrals over $\Omega_c \setminus \Omega_0$ and $\Omega_0 \setminus \Omega_c$. The functional in Eq. (7.9) consists of seven terms which we group into four terms for each of which we perform the ϵ-expansion:

(1)

$$\gamma \int_{\Omega_c} d\Omega \left[s(u, \nabla_a u) - R_0^2 \right] + \gamma R_0^2 \left(\int_{\Omega_c \setminus \Omega_0} d\Omega - \int_{\Omega_0 \setminus \Omega_c} d\Omega \right)$$

$$= \gamma R_0^2 \int_{\Omega_0} d\Omega \left[(1 + \epsilon v)^2 + \frac{\epsilon^2}{2} (\nabla_a v)^2 - 1 \right]$$

$$+ \gamma R_0^2 \left(\int_{\Omega_c \setminus \Omega_0} d\Omega - \int_{\Omega_0 \setminus \Omega_c} d\Omega \right) \left[1 + 2\epsilon v \right] + O(\epsilon^3)$$

$$= \gamma R_0^2 \int_{\Omega_0} d\Omega \left[2\epsilon v + \epsilon^2 v^2 + \frac{\epsilon^2}{2} (\nabla_a v)^2 \right]$$

$$+ \gamma R_0^2 \int_0^{2\pi} d\phi \int_{\theta_0}^{\theta_c(\phi)} d\theta \sin\theta \left[1 + 2\epsilon v \right] + O(\epsilon^3), \quad \text{(B.1)}$$

APPENDIX B. DERIVATION OF THE FREE ENERGY FUNCTIONAL

(2)

$$
\begin{aligned}
&- \gamma R_0^2 \frac{\cos\theta_0}{2} \int_0^{2\pi} d\phi \left[(1+\epsilon v_c)^2 \sin^2\theta_c - \sin^2\theta_0\right] \\
&= -\gamma R_0^2 \frac{\cos\theta_0}{2} \int_0^{2\pi} d\phi \left[(1+\epsilon v_c)^2 \left(1 - \frac{\cos^2\theta_0}{(1+\epsilon v_c)^2}\right) - \sin^2\theta_0\right] \\
&= -\gamma R_0^2 \frac{\cos\theta_0}{2} \int_0^{2\pi} d\phi \left[(1+\epsilon v_c)^2 - \cos^2\theta_0 - \sin^2\theta_0\right] \\
&= -\gamma R_0^2 \cos\theta_0 \int_0^{2\pi} d\phi \left[\epsilon v_c + \frac{\epsilon^2 v_c^2}{2}\right] + O(\epsilon^3), \quad \text{(B.2)}
\end{aligned}
$$

where we skipped the explicit dependence of v_c and θ_c on ϕ and in the first equality we used

$$\cos\theta_c = \frac{\cos\theta_0}{1+\epsilon v_c}, \quad \text{(B.3)}$$

which follows from the analysis of small perturbations of a spherical cap.

(3)

$$
\begin{aligned}
&-\frac{1}{3}\int_{\Delta\Omega} d\Omega\, \Pi(\Omega)\left[(R_0+u)^3 - R_0^3\right] - \frac{\lambda}{3}\int_{\Omega_c} d\Omega \left[(R_0+u)^3 - R_0^3\right] \\
&= -\frac{\gamma R_0^2}{3} \int_{\Omega_0} d\Omega \left[\epsilon\pi(\Omega) + 2 + \epsilon\mu\right]\left[(1+\epsilon v)^3 - 1\right] \\
&\quad - \frac{\gamma R_0^2}{3}\left(\int_{\Omega_c\setminus\Omega_0} d\Omega - \int_{\Omega_0\setminus\Omega_c} d\Omega\right) 6\epsilon v + O(\epsilon^3) \\
&= -\gamma R_0^2 \int_{\Omega_0} d\Omega \left[2\epsilon v + 2\epsilon^2 v^2 + \epsilon^2\big(\pi(\Omega)+\mu\big)v\right] \\
&\quad - \gamma R_0^2 \int_0^{2\pi} d\phi \int_{\theta_0}^{\theta_c(\phi)} d\theta\, \sin\theta\, 2\epsilon v + O(\epsilon^3), \quad \text{(B.4)}
\end{aligned}
$$

(4)

$$\frac{\lambda}{3}\left(\int_{\Omega_c\setminus\Omega_0} d\Omega - \int_{\Omega_0\setminus\Omega_c} d\Omega\right)\left[(R_s(\theta))^3 - R_0^3\right]$$

$$= \frac{\gamma R_0^2}{3}(2+\epsilon\mu)\int_0^{2\pi} d\phi, \int_{\theta_0}^{\theta_c(\phi)} d\theta \sin\theta \left(\frac{\cos^3\theta_0}{\cos^3\theta} - 1\right)$$

$$= \frac{\gamma R_0^2}{3}(2+\epsilon\mu)\int_0^{2\pi} d\phi \int_{\cos\theta_c(\phi)}^{\cos\theta_0} dx \left[\frac{\cos^3\theta_0}{x^3} - 1\right]$$

$$= \frac{\gamma R_0^2}{3}(2+\epsilon\mu)\int_0^{2\pi} d\phi \left[-\frac{\cos^3\theta_0}{2}\left(\frac{1}{\cos^2\theta_0} - \frac{1}{\cos^2\theta_c}\right) - (\cos\theta_0 - \cos\theta_c)\right]$$

$$= \frac{\gamma R_0^2}{3}(2+\epsilon\mu)\int_0^{2\pi} d\phi\,(\cos\theta_0 - \cos\theta_c)\left[\frac{\cos\theta_0(\cos\theta_0 + \cos\theta_c)}{2\cos^2\theta_c} - 1\right]$$

$$= \frac{\gamma R_0^2}{3}(2+\epsilon\mu)\int_0^{2\pi} d\phi\,\cos\theta_0 \epsilon v_c \left[\frac{1}{2}(2+\epsilon v_c)(1+\epsilon v_c) - 1\right]$$

$$= \gamma R_0^2 \int_0^{2\pi} d\phi\,\cos\theta_0 \epsilon^2 v_c^2 + O(\epsilon^3), \quad (B.5)$$

where in the last but one equality we again used Eq. (B.3).

Summing up all the terms (we neglect the term $-\lambda\delta V$) the resulting free energy functional $\mathcal{F}$ can be splitted into two parts

$$\mathcal{F} = \mathcal{F}_{surf} + \mathcal{F}_{bc}, \quad (B.6)$$

where

$$\frac{1}{\gamma R_0^2}\mathcal{F}_{surf}[\{v(\Omega)\}] = \epsilon^2 \int_{\Omega_0} d\Omega \left[\frac{1}{2}(\nabla_a v)^2 - v^2 - \left(\pi(\Omega) + \mu\right)v\right] + O(\epsilon^3) \quad (B.7)$$

is the *surf*ace contribution and

$$\frac{1}{\gamma R_0^2}\mathcal{F}_{bc}[\{v(\Omega)\}]$$

$$= \int_0^{2\pi} d\phi \left(\int_{\theta_0}^{\theta_c(\phi)} d\theta \sin\theta[1+2\epsilon v - 2\epsilon v] - \cos\theta_0\left[\epsilon v_c + \frac{\epsilon^2 v_c^2}{2}\right] + \cos\theta_0 \epsilon^2 v_c^2\right) + O(\epsilon^3)$$

$$= \int_0^{2\pi} d\phi \left(\cos\theta_0 - \cos\theta_c - \cos\theta_0\left[\epsilon v_c + \frac{\epsilon^2 v_c^2}{2}\right] + \cos\theta_0 \epsilon^2 v_c^2\right) + O(\epsilon^3)$$

$$= \cos\theta_0 \int_0^{2\pi} d\phi \left(\epsilon v_c - \epsilon^2 v_c^2 - \epsilon v_c - \frac{\epsilon^2 v_c^2}{2} + \epsilon^2 v_c^2\right) + O(\epsilon^3)$$

$$= -\frac{\epsilon^2 \cos\theta_0}{2}\int_0^{2\pi} d\phi\,v_c^2 + O(\epsilon^3) = -\frac{\epsilon^2 \cos\theta_0}{2}\int_0^{2\pi} d\phi\,(v|_{\theta_0})^2 + O(\epsilon^3) \quad (B.8)$$

is the *b*oundary *c*ontribution. In the third equality we have used Eq. (B.3) and in the last equality we have replaced v_c by $v|_{\theta_0}$ which gives a correction $O(\epsilon^3)$.

Finally, we also note that taking into account the constraint of a fixed center of mass would yield a contribution $-f_{CM}(x_{CM} - x_{CM,ref}) = O(\epsilon^2)$ to $\mathcal{F}$, but the corresponding contribution to $\mathcal{F}|_{bc}$ would be $O(\epsilon^3)$ (one ϵ due to f_{CM}, one due to v and one due to $\theta_c(\phi) - \theta_0$) and therefore we have omitted it here.

Appendix C

Local approximation of the Green's equation for a sessile droplet

The Green's equations Eq. (7.44) and Eq. (7.45) can be written shortly as

$$-(\nabla_a^2 + 2)G_\sigma(\Omega, \Omega') = \delta(\Omega, \Omega') + \Delta_\sigma(\Omega, \Omega'), \tag{C.1}$$

where the function Δ_σ denotes all the regular non-homogenous terms on the rhs.

We study the behavior of Eq. (C.1) in the case when $\Omega \to \Omega'$. We choose new spherical coordinates $\tilde{\Omega} = (\tilde{\theta}, \tilde{\phi})$ defined with respect to the axis $\tilde{z} = \hat{R}z$, where $\hat{R}$ is a rotation operator and thus $\tilde{\Omega} = \hat{R}^{-1}\Omega$ (see Fig. C.1). The Green's equation (Eq. (C.1)) can be then written as

$$-(\tilde{\nabla}_a^2 + 2)\tilde{G}_\sigma(\tilde{\Omega}, \tilde{\Omega}') = \delta(\tilde{\Omega}, \tilde{\Omega}') + \tilde{\Delta}_\sigma(\tilde{\Omega}, \tilde{\Omega}'), \tag{C.2}$$

where

$$\tilde{\nabla}_a := \boldsymbol{e}_{\tilde{\theta}}\partial_{\tilde{\theta}} + \frac{\boldsymbol{e}_{\tilde{\phi}}}{\sin\tilde{\theta}}\partial_{\tilde{\phi}} \tag{C.3}$$

and the functions $\tilde{G}_\sigma$ and $\tilde{\Delta}_\sigma$ are defined as $\tilde{G}_\sigma(\tilde{\Omega}, \tilde{\Omega}') := G_\sigma(\hat{R}\tilde{\Omega}, \hat{R}\tilde{\Omega}')$ and $\tilde{\Delta}_\sigma(\tilde{\Omega}, \tilde{\Omega}') := \Delta_\sigma(\hat{R}\tilde{\Omega}, \hat{R}\tilde{\Omega}')$. Moreover, we have used the fact that the delta function is invariant with respect to rotations, so that $\delta(\hat{R}\tilde{\Omega}, \hat{R}\tilde{\Omega}') = \delta(\tilde{\Omega}, \tilde{\Omega}')$.

Next, we define the mapping $(\tilde{\theta}, \tilde{\phi}) \mapsto (\tilde{\rho}, \tilde{\phi})$ which associates each point on the sphere in the neighborhood of the $\tilde{z}$-axis with its projection onto the tangent plane (defined at the point of intersection of the $\tilde{z}$-axis with the sphere), where $(\tilde{\rho}, \tilde{\phi})$ are polar coordinates in the tangent plane. Thus, the relation between $\tilde{\rho}$ and $\tilde{\theta}$ is

$$\tilde{\rho} = R_0 \sin\tilde{\theta}. \tag{C.4}$$

For simplicity of the notation we skip the tilde so that $(\tilde{\rho}, \tilde{\phi}) \equiv (\rho, \phi)$. The Laplace-

134 APPENDIX C. LOCAL APPROXIMATION OF THE GREEN'S EQUATION

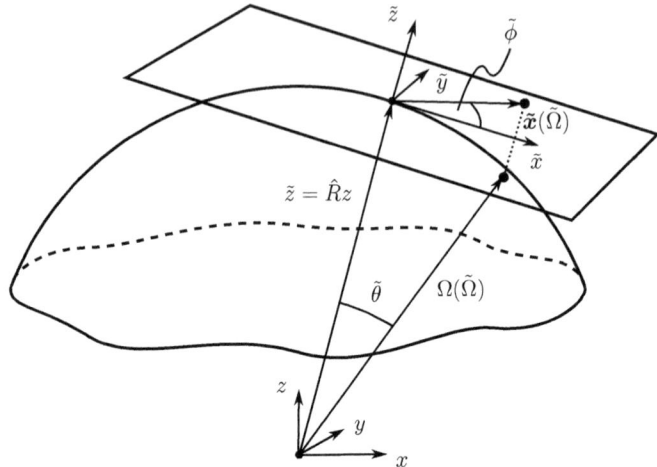

Figure C.1: Tangent plane and local flat coordinates at a unit sphere. Projection of the point associated with the direction Ω (or, equivalently, $\tilde{\Omega}$) onto the $\tilde{x}\tilde{y}$-plane defines $\tilde{\boldsymbol{x}}(\tilde{\Omega})$.

Beltrami operator ∇_a^2 can be written in terms of the polar coordinates as

$$\nabla_a^2 = \frac{1}{\sin\theta}\partial_\theta \sin\theta\, \partial_\theta + \frac{1}{\sin^2\theta}\partial_\phi^2 = \partial_\theta^2 + \cot\theta\, \partial_\theta + \frac{1}{\sin^2\theta}\partial_\phi^2$$

$$= \frac{d\rho}{d\theta}\partial_\rho \frac{d\rho}{d\theta}\partial_\rho + \frac{\sqrt{R_0^2 - \rho^2}}{\rho}\frac{d\rho}{d\theta}\partial_\rho + \frac{R_0^2}{\rho^2}\partial_\phi^2$$

$$= \sqrt{R_0^2 - \rho^2}\,\partial_\rho\sqrt{R_0^2 - \rho^2}\,\partial_\rho + \frac{R_0^2 - \rho^2}{\rho}\partial_\rho + \frac{R_0^2}{\rho^2}\partial_\phi^2$$

$$= (R_0^2 - \rho^2)\partial\rho^2 + \frac{1}{\rho}(R_0^2 - 2\rho^2)\,\partial_\rho + \frac{R_0^2}{\rho^2}\partial_\phi^2$$

$$= R_0^2\left[\frac{1}{\rho}\partial_\rho\, \rho\, \partial_\rho + \frac{1}{\rho^2}\partial_\phi^2\right] - \partial_\rho\, \rho^2\, \partial_\rho = R_0^2\nabla_\parallel^2 - \partial_\rho\, \rho^2\, \partial_\rho, \quad \text{(C.5)}$$

where $\nabla_\parallel^2$ is the two-dimensional Laplace operator on the tangent plane. The delta function transforms as (here also skipping the tilde)

$$\delta(\Omega, \Omega') = \frac{1}{\sin\theta}\delta(\theta - \theta')\,\delta(\phi - \phi') = \frac{R_0}{\rho}\delta(\theta(\rho) - \theta'(\rho'))\,\delta(\phi - \phi')$$

$$= \frac{R_0}{|d\theta/d\rho|}\frac{1}{\rho}\delta(\rho - \rho')\,\delta(\phi - \phi') = R_0\sqrt{R_0^2 - \rho^2}\,\delta(\boldsymbol{x} - \boldsymbol{x}'), \quad \text{(C.6)}$$

where we have used $(d\theta/d\rho)^{-1} = d\rho/d\theta = \sqrt{R_0^2 - \rho^2}$. Finally, Eqs. (C.2), (C.5) and

(C.6) lead to the Green's equation expressed in terms of the local "flat" coordinates (Eq. (7.54)).

Appendix D

Calculation of the functions $H(\Omega')$ and $I(\Omega')$ in the Green's function $G_B(\Omega, \Omega')$

D.1 Derivation of the function $I(\Omega')$

Imposing the boundary condition

$$G_B(\Omega, \Omega')|_{\Omega=(\theta=\pi/2,\phi)} = 0, \tag{D.1}$$

on the Green's function in Eq. (7.67) yields

$$2G(\bar{\theta} = \pi/2)\cos\theta' + I(\theta') = 0, \tag{D.2}$$

from which it follows that

$$I(x) = \frac{\cos x}{4\pi}. \tag{D.3}$$

D.2 Derivation of the function $H(\Omega')$

In order to calculate the function $H(\Omega')$ we use the volume constraint

$$\int_0^{2\pi} d\phi \int_0^{\pi/2} d\theta \sin\theta \, G_B(\Omega, \Omega') = 0, \tag{D.4}$$

which together with Eq. (7.67) yields

$$C(\Omega') - B(\Omega') + 2A\cos\theta' + \pi H(\Omega') + 2\pi I(\Omega') = 0, \tag{D.5}$$

where we have defined

$$A := \int_0^{2\pi} d\phi \int_0^{\pi/2} d\theta \sin\theta \, G(\Omega, \Omega_\pi), \tag{D.6}$$

$$B(\Omega') = B(\theta') := \int_0^{2\pi} d\phi \int_0^{\pi/2} d\theta \sin\theta \, G(\Omega, \hat{Z}\Omega'), \tag{D.7}$$

$$C(\Omega') = C(\theta') := \int_0^{2\pi} d\phi \int_0^{\pi/2} d\theta \sin\theta \, G(\Omega, \Omega'). \tag{D.8}$$

APPENDIX D. CALCULATION OF THE FUNCTIONS $H(\Omega')$ AND $I(\Omega')$

From Eq. (D.5) we obtain $H(\Omega') \equiv H(\theta')$ and

$$H(\theta') = \frac{1}{\pi}\left[B(\theta') - C(\theta')\right] - \frac{2A}{\pi}\cos\theta' - 2I(\theta'). \tag{D.9}$$

The coefficient A, due to the axial symmetry of the integrand, can be carried out immediately with the result

$$A = \frac{5}{24} - \frac{\ln 2}{4}. \tag{D.10}$$

The volume constraint for G in the form $\int_{S_2} d\Omega\, G(\Omega, \Omega') = 0$ yields

$$B(\theta') + C(\theta') = 0. \tag{D.11}$$

Therefore, we have $C(\theta') - B(\theta') = -2B(\theta')$, and thus only the function $B(\theta')$ is left to be calculated in Eq. (D.9). By using the symmetry of the Green's function $G(\Omega, \Omega') = G(\Omega', \Omega)$ and the explicit form of G in Eq. (6.26) we can write

$$B(\theta') = \int_0^{2\pi} d\phi \int_0^{\pi/2} d\theta \sin\theta\, G(\hat{Z}\Omega, \Omega') = \int_0^{2\pi} d\phi \int_{\pi/2}^{\pi} d\theta \sin\theta\, G(\Omega, \Omega') =$$

$$= -\frac{1}{4\pi}\int_0^{2\pi} d\phi \int_{\pi/2}^{\pi} d\theta \sin\theta \left[\frac{1}{2} + \frac{4}{3}\cos\bar{\theta} + \cos\bar{\theta}\ln\left(\frac{1 - \cos\bar{\theta}}{2}\right)\right], \tag{D.12}$$

where

$$\cos\bar{\theta} = \cos\theta\cos\theta' + \sin\theta\sin\theta'\cos\phi. \tag{D.13}$$

We introduce the following auxiliary integrals

$$J_0(y) := \frac{1}{\pi}\int_0^{\pi} d\phi \ln(1 - y\cos\phi), \tag{D.14}$$

$$J_1(y) := \frac{1}{\pi}\int_0^{\pi} d\phi \cos\phi \ln(1 - y\cos\phi). \tag{D.15}$$

First, we calculate

$$\frac{dJ_0}{dy} = \frac{1}{\pi}\int_0^{\pi} d\phi \frac{-\cos\phi}{1 - y\cos\phi} = \frac{1}{y} - \frac{1}{y\sqrt{1-y^2}}. \tag{D.16}$$

Then, $J_0(y)$ can be obtained by integration with respect to y,

$$J_0(y) = \int_0^y dy'\left(\frac{1}{y'} - \frac{1}{y'\sqrt{1-y'^2}}\right) = \ln\left(\frac{\sqrt{1-y^2}+1}{2}\right). \tag{D.17}$$

Analogically, we obtain

$$J_1(y) = \frac{\sqrt{1-y^2}-1}{y}. \tag{D.18}$$

Denoting $s := \sin\theta'$, $c := \cos\theta'$, $x := -\cos\theta$ and $y(x) := s\sqrt{1-x^2}/(1+cx)$, and performing the integration over ϕ, we can rewrite Eq. (D.12) in the form

$$B = -\frac{1}{2}\int_0^1 dx \left[\frac{1}{2} - \frac{4}{3}cx - cx\ln\left(\frac{1+cx}{2}\right) - cxJ_0\bigl(y(x)\bigr) + s\sqrt{1-x^2}J_1\bigl(y(x)\bigr)\right]. \tag{D.19}$$

D.2. DERIVATION OF THE FUNCTION $H(\Omega')$

Performing the following simplifications

$$-cx \ln\left(\frac{1+cx}{2}\right) - cx J_0\big(y(x)\big) = -cx \ln \frac{(1+c)(1+x)}{4}, \quad \text{(D.20)}$$

$$s\sqrt{1-x^2} J_1\big(y(x)\big) = -(1-c)(1-x), \quad \text{(D.21)}$$

we obtain

$$B(\theta') = -\frac{1}{2}\int_0^1 dx \left[c - \frac{1}{2} + x\left(1 - \frac{7c}{3} - c\ln\frac{1+c}{4}\right) - cx\ln(1+x)\right] =$$
$$= \frac{\cos\theta'}{4}\left(\frac{5}{6} + \ln\frac{1+\cos\theta'}{4}\right). \quad \text{(D.22)}$$

It can be checked, that for $\theta' = 0$ the relation $C(\theta' = 0) = A$ is recovered. Finally, Eq. (D.5) yields expression for $H(x)$ in the form

$$H(x) = \frac{1}{2\pi}\left[\cos x \ln\left(\frac{1+\cos x}{2}\right) - \cos x\right]. \quad \text{(D.23)}$$

Appendix E

Exact results and linear theory in the axisymmetric case for arbitrary θ_0

E.1 Exact results

Consider a spherical particle of radius a positioned at the apex of a sessile droplet, characterized by a fixed contact angle θ_0 at the substrate, which means that the contact line at the substrate is not pinned and can move freely (model A). Due to axial symmetry and in absence of gravity the free energy can be calculated analytically solving the full non-linear Young-Laplace equation. For simplicity, we assume that the interface has no overhangs so that the shape of the droplet can be described by $z = z(r)$, which is particularly the case for $\theta_0 < \pi/2$ and $\theta_p = \pi/2$ independently of the position of the contact line at the particle. A similar analysis can be performed also for the case of a pinned contact line at the substrate, in which the surface energy associated with the substrate is constant. The situation with a free contact line is actually more complicated, because in this case the surface free energy associated with the substrate can vary.

E.1.1 Free energy functional

In the following calculations we take a as the unit of length and γa^2 as the unit of energy. The free energy can be expressed as

$$\tilde{F}^\star_{A0}(h, \theta_0, \theta_p, R_0, \lambda) = \min_{\{z(r)\}} \tilde{\mathcal{F}}[\{z(r)\}], \tag{E.1}$$

where the index $A0$ indicates the free contact line at the substrate and the axisymmetric configuration, respectively. The free energy functional $\tilde{\mathcal{F}}$ reads

$$\tilde{\mathcal{F}}[\{z(r)\}; h, \theta_0, \theta_p, V_l, \lambda] = S_{lg} - S_{lg,ref} - (S_{ol} - S_{ol,ref})\cos\theta_0 \\ - (S_{pl} - S_{pl,ref})\cos\theta_p - \lambda(V - V_l), \tag{E.2}$$

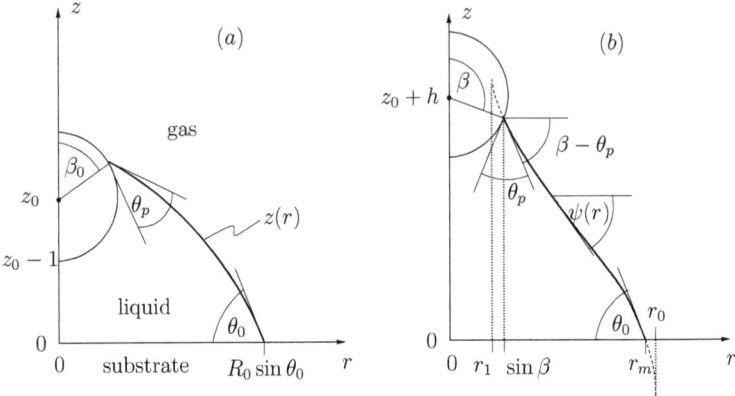

Figure E.1: Cross section of an axially symmetric configuration of a droplet and a particle; (a) reference configuration for $f = 0$ and $h = 0$, which leads to a liquid-gas interface with the shape of a spherical cap and the contact line at the particle at β_0; (b) sketch of a configuration for an arbitrary β corresponding to $f \neq 0$ and $h \neq 0$. In the studied model θ_0 is kept fixed and $r_m \neq R_0 \sin \theta_0$. The dashed lines denote continuations of the analytical profile corresponding to an unduloid or a nodoid beyond the physical regime bounded by $r = \sin \beta$ and $r = r_m$ and terminating at maximal and minimal radii r_0 and r_1, respectively.

with the volume V_l of liquid being a function of the contact angles θ_0 and θ_p, and of the droplet radius R_0 in the reference configuration, i.e., $V_l = V_l(\theta_0, \theta_p, R_0)$. The Lagrange multiplier λ can be determined from the condition $V = V_l$, which renders $\lambda = \lambda(h, \theta_0, \theta_p, R_0)$ and which upon insertion into $\tilde{F}^*_{A0}(h, \theta_0, \theta_p, R_0, \lambda)$ yields the free energy $\tilde{F}_{A0}(h, \theta_0, \theta_p, R_0)$.

It is convenient to express the contact areas S_{0l}, S_{pl}, and S_{lg} in terms of auxiliary variables: the angle β describing the position of the contact line at the particle and the radius r_m of the circle formed by the contact line at the substrate (see Fig. E.1(b)). One can write

$$S_{0l} = \pi r_m^2, \tag{E.3}$$
$$S_{pl} = 2\pi(1 + \cos \beta) = 4\pi \cos^2(\beta/2), \tag{E.4}$$
$$S_{lg} = 2\pi \int_{\sin \beta}^{r_m} dr\, r\sqrt{1 + \dot{z}^2}, \tag{E.5}$$

with $\dot{z} = dz(r)/dr$. The volume of liquid V can be expressed as

$$V = V_{lg} + \pi(\sin^2 \beta) z(r = \sin \beta) - \frac{\pi}{3}(1 + \cos \beta)^2 (2 - \cos \beta), \tag{E.6}$$

with

$$V_{lg} = 2\pi \int_{\sin \beta}^{r_m} dr\, r z(r), \tag{E.7}$$

E.1. EXACT RESULTS

where V_{lg} is the volume enclosed between the substrate and the liquid-gas interface, the second term in Eq. (E.6) is the volume between the substrate and the circular disc determined by the contact line at the particle, and the third term is the volume of that part of the particle which is immersed in the liquid.

The equilibrium profile $z(r)$, which minimizes $\tilde{\mathcal{F}}$ obeys the Young-Laplace equation in cylindrical coordinates (see,e.g., Langbein 2002):

$$\frac{1}{r}\frac{d}{dr}\frac{r\dot{z}}{\sqrt{1+\dot{z}^2}} = -\lambda \tag{E.8}$$

The boundary conditions are determined by Young's law at the particle and at the substrate (which follow from the condition of vanishing of the variation of $\tilde{\mathcal{F}}$ at the boundaries),

$$\psi(r = \sin\beta) = \beta - \theta_p, \tag{E.9}$$
$$\psi(r = r_m) = \theta_0, \tag{E.10}$$

where $\psi = \psi(r)$ is the angle between the r-axis and the tangent of the profile $z(r)$ (see Fig. E.1(b)) defined by

$$\sin\psi(r) := -\frac{\dot{z}}{\sqrt{1+\dot{z}^2}}. \tag{E.11}$$

The first integral of Eq. (E.8) reads

$$\frac{\dot{z}}{\sqrt{1+\dot{z}^2}} \equiv -\sin\psi(r) = -\frac{\lambda r}{2} + \frac{c}{r}, \tag{E.12}$$

where c is an integration constant. When evaluated at the boundaries and using Eqs. (E.9) and (E.10) this leads to the following set of equations:

$$-\sin(\beta - \theta_p) = -\frac{\lambda\sin\beta}{2} + \frac{c}{\sin\beta}, \tag{E.13}$$

$$-\sin\theta_0 = -\frac{\lambda r_m}{2} + \frac{c}{r_m}. \tag{E.14}$$

Solving with respect to λ and c one obtains

$$\lambda = \frac{2[r_m\sin\theta_0 - \sin\beta\sin(\beta-\theta_p)]}{r_m^2 - \sin^2\beta} = \lambda(r_m, \beta), \tag{E.15}$$

$$c = -\frac{r_m^2\sin\beta\sin(\beta-\theta_p) - r_m\sin\theta_0\sin^2\beta}{r_m^2 - \sin^2\beta} = c(r_m, \beta). \tag{E.16}$$

E.1.2 Interface profile, surface area and volume

The equilibrium profile $z(r)$ is obtained by solving Eq. (E.12) for $\dot{z}$ and by subsequently integrating:

$$z(r) = -\frac{1}{\lambda}\int_{\sin\beta}^{r} dr \frac{\lambda r^2 - 2c}{\sqrt{(r^2-r_1^2)(r_0^2-r^2)}} =$$
$$= r_0\big[E(\phi, q) - E(\phi_2, q)\big] - \frac{2c}{\lambda r_0}\big[F(\phi, q) - F(\phi_2, q)\big]. \tag{E.17}$$

where $F(\phi, q)$ and $E(\phi, q)$ are the incomplete elliptic integrals of the first and second kind, respectively (Prudnikov et al. 1986b). The integral in Eq. (E.5) can be evaluated as

$$S_{lg} = \frac{4\pi}{|\lambda|} \int_{\sin\beta}^{r_m} dr \frac{r^2}{\sqrt{(r^2 - r_1^2)(r_0^2 - r^2)}} =$$
$$= 4\pi \frac{r_0}{|\lambda|} \left(E(q) - E(\phi_1, q) - E(\phi_2, q) + \frac{1}{r_0 \sin\beta} \sqrt{(r_0^2 - \sin^2\beta)(\sin^2\beta - r_1^2)} \right). \quad (E.18)$$

The expression in Eq. (E.7) can be evaluated by integrating by parts, which leads to

$$V_{lg} = \frac{\pi}{\lambda} \int_{\sin\beta}^{r_m} dr \frac{\lambda r^4 - 2cr^2}{\sqrt{(r^2 - r_1^2)(r_0^2 - r^2)}} - \pi(\sin^2\beta) z(r = \sin\beta) =$$
$$= \pi r_0 \left(\kappa \Big[E(q) - E(\phi_1, q) - E(\phi_2, q) \Big] - \frac{r_1^2}{3} \Big[K(q) - F(\phi_1, q) - F(\phi_2, q) \Big] \right.$$
$$+ \left[\frac{\sin\beta}{3} + \frac{\kappa}{\sin\beta} \right] \sqrt{(r_0^2 - \sin^2\beta)(\sin^2\beta - r_1^2)} - \frac{r_m}{3} \sqrt{(r_0^2 - r_m^2)(r_m^2 - r_1^2)} \right)$$
$$- \pi(\sin^2\beta) z(r = \sin\beta). \quad (E.19)$$

where in Eqs. (E.18) and (E.19) $K(q)$ and $E(q)$ are the complete elliptic integrals of the first and second kind, respectively. Moreover, one has

$$\sin\phi = \sqrt{\frac{r_0^2 - r^2}{r_0^2 - r_1^2}} \quad (E.20)$$

$$q = \sqrt{1 - \frac{r_1^2}{r_0^2}}, \quad (E.21)$$

$$\kappa = \frac{2c}{3\lambda} + \frac{8}{3\lambda^2}, \quad (E.22)$$

$$\sin\phi_1 = \frac{r_0}{\sin\beta} \sqrt{\frac{\sin^2\beta - r_1^2}{r_0^2 - r_1^2}}, \quad (E.23)$$

and

$$\sin\phi_2 = \sqrt{\frac{r_0^2 - r_m^2}{r_0^2 - r_1^2}}. \quad (E.24)$$

The shape of the droplet is always a section of a nodoid or an unduloid (Langbein 2002) characterized by a maximal and a minimal radius r_0 and r_1 (see Fig. E.1(b)), respectively, which are two distinct solutions of the equation $\dot{z}(r) = -\infty$ and are given by

$$r_0 = \frac{1}{\lambda} \left(\sqrt{1 + 2\lambda c} + 1 \right), \quad (E.25)$$

$$r_1 = \frac{1}{\lambda} \left(\sqrt{1 + 2\lambda c} - 1 \right). \quad (E.26)$$

E.1. EXACT RESULTS

So far the interface profile (Eq. (E.17)) and the free energy (Eqs. (E.1)-(E.4) and (E.18)) are determined in terms of the independent variables r_m and β (through Eqs. (E.15), (E.16) and (E.22)-(E.26)). One can replace them by the physically more directly accessible variables h and V_l. To this end we note that the vertical displacement h of the particle is determined by (see Fig. E.1(b))

$$z_0 + h = z(r = \sin\beta) - \cos\beta, \tag{E.27}$$

where $z_0 = z_0(\theta_0, \theta_p, R_0)$ is the height of the particle center above the substrate in the reference configuration with the droplet shape given by the cap of a sphere (see, c.f., Eq. (E.34) and Fig. E.2). This renders the set of equations

$$h = h(r_m, \beta, V_l), \tag{E.28}$$
$$V_l = V(r_m, \beta), \tag{E.29}$$

where $h(r_m, \beta, V_l)$ is given by Eq. (E.27) (with the dependence on r_m entering via $z(r)$) and $V(r_m, \beta)$ is defined (suppressing the explicit dependence on θ_0 and θ_p) via Eqs. (E.6) and (E.19) together with (E.15), (E.16) and (E.22)-(E.26). Equations (E.28) and (E.29) provide implicitly the maximal radius r_m and the angle β as functions of V_l and h.

Finally, with keeping in mind that here all length scales are measured in units of a, the constant liquid volume V_l can be expressed in terms of the materials parameters θ_0 and θ_p as well as the drop size R_0. As one can infer from the geometrical features shown in Fig. E.2, V_l is given by

$$V_l(R_0, \theta_0, \theta_p) = \frac{4\pi}{3}\left(\left[f_0(\theta_0) - f_0(\beta_0 - \theta_p)\right]R_0^3 - f_0(\pi - \beta_0)\right) \tag{E.30}$$

with

$$\sin\beta_0 = R_0 \sin(\beta_0 - \theta_p) \tag{E.31}$$

so that

$$\beta_0(R_0, \theta_p) = \arcsin\left((R_0 \sin\theta_p)/(R_0^2 - 2R_0 \cos\theta_p + 1)^{1/2}\right), \tag{E.32}$$

and where the function $f_0(\theta_0)$ expresses the volume of a unit spherical cap characterized by the polar angle θ_0 as a fraction of the volume $4\pi/3$ of a unit sphere:

$$f_0(x) := (2 + \cos x)\sin^4(x/2). \tag{E.33}$$

Additionally, since the presence of the particle does not change the contact angle θ_0, the liquid volume V_l can be expressed in terms of the size $\bar{R}_0 = [3V_l/(4\pi f_0(\theta_0))]^{1/3}$ of the droplet without the particle. The vertical position z_0 of the particle in the reference configuration can be expressed as

$$z_0(R_0, \theta_0, \theta_p) = R_0 \cos(\beta_0 - \theta_p) - R_0 \cos\theta_0 - \cos\beta_0. \tag{E.34}$$

146 APPENDIX E. EXACT RESULTS AND LINEAR THEORY FOR ARBITRARY θ_0

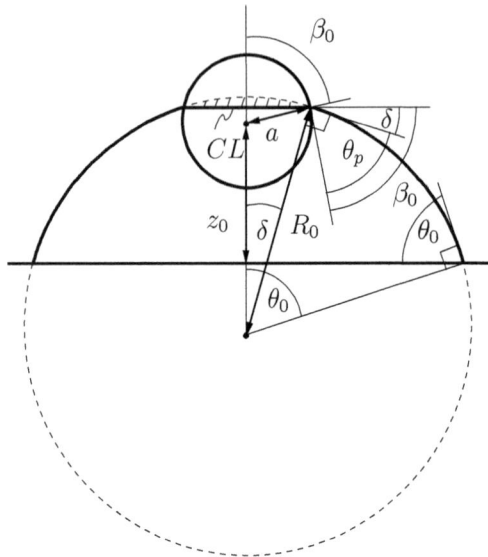

Figure E.2: Sketch of the reference configuration. The angle β_0 can be found by comparing two expressions for the radius of the contact line at the particle, namely $a \sin \beta_0 = R_0 \sin \delta$, where δ can be found equal $\beta_0 - \theta_p$ (which then yields Eq. (E.31)). In calculating the volume V_l of the liquid one must subtract from the volume of the spherical cap of radius R_0 the volume of the immersed part of the particle (beneath the contact line CL) and additionally the small spherical cap indicated in the picture as a hatched region (see Eq. (E.30)).

E.1.3 Free energy and mean-force

The sets of equations (E.28) - (E.34) (which must be solved numerically) implicitly yield $\beta = \beta(h, R_0)$ and $r_m = r_m(h, R_0)$, which can be substituted into the expressions for the surface areas in Eqs. (E.3), (E.4), and (E.18). Finally, the free energy functional $\tilde{\mathcal{F}}$ given as a combination of these areas (Eq. (E.2)) yields the free energy $\tilde{F}_{A0}(h, \theta_0, \theta_p, R_0)$.

The results of the calculations for $\theta_0 = \pi/3, \theta_p = \pi/2$, and $V_l = 79 \times (4\pi/3)$ are presented in Fig. E.3(a) (for the results concerning the droplet shape see Fig. E.5). One can distinguish seven branches of the free energy both for $h < 0$ and for $h > 0$. Branch 1 corresponds to the globally stable, full analytic solution of the Young-Laplace equation. Branch 2 is the free energy under the constraint that the droplet keeps the shape of a spherical cap, which is the equilibrium configuration for $h = 0$. For $h \neq 0$ this constraint induces θ_p to deviate from its value $\pi/2$. Lifting this constraint lowers the free energy towards the globally stable branch 1, on which θ_p has its fixed value $\pi/2$. For $h/a > 0.93$ and $h/a < -1.04$ the particle detaches from the interface if one keeps the droplet shape to be a spherical cap (3, 4). On branch 3 (4) this detached

E.1. EXACT RESULTS

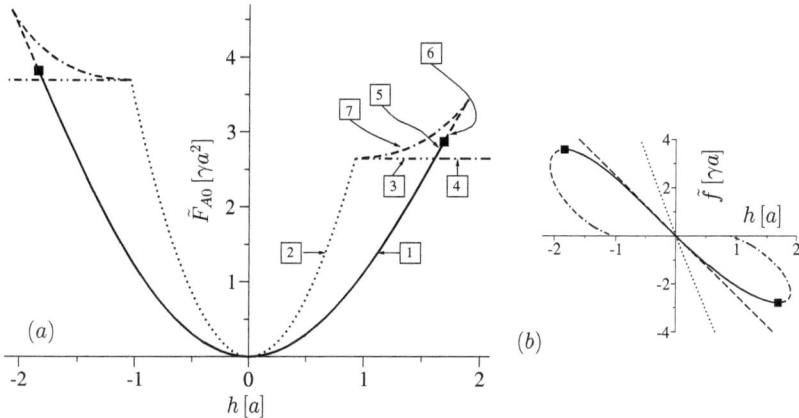

Figure E.3: The surface free energy $\tilde{F}_{A0}(h)$ (a) for axially symmetric configurations ($\alpha = 0$) with $\theta_0 = \pi/3, \theta_p = \pi/2$, and $V_l = 79\frac{4\pi}{3}a^3$ and the total mean force $\tilde{f}(h) = -d\tilde{F}_{A0}/dh$ (b), see Eq. (E.35), as functions of immersion h. For $h > 0$ and $h < 0$ one can distinguish 7 branches (see main text). For reasons of clarity only the branches for $h > 0$ have been tagged in the plot. The various line codes in (a) and (b) correspond to each other. In (b) the straight long-dashed line indicates the slope of the analytic solution at $h = 0$.

configuration is metastable (globally stable). The extension of branch 1 beyond the horizontal branches 3 and 4 is metastable on branch 5 up to the filled square, which indicates an inflection point. Beyond that point branch 6 is unstable ($\tilde{F}(h)$ is convex). Branch 7 is an extension of branch 2 which is metastable with respect to detachment, i. e., branches 3 and 4. Branches 6 and 7 merge at the point at which the unstable solution along branch 6 ceases to exist. The equilibrium configurations are given by the branches 1 and 4.

We note that in the above discussion "stability" *of the branches of* $\tilde{F}(h)$ should be understood as follows: the stable and the metastable branches are convex; the stable branch has a lower energy than the metastable branch; the unstable branch is concave. On the other hand the stability *of configurations of the particle* refers to the situations in the presence of an external force f fixing the position of the particle at a given height h. In such a case the particle experiences an effective potential $\tilde{F}(h) - fh$, whose global or local (non-global) minima correspond to the equilibrium stable or metastable configurations, respectively. The external force f must counterbalance the mean-force $\tilde{f}$, which might be expressed as

$$\tilde{f}(h) = -\frac{d\tilde{F}_{A0}}{dh}. \tag{E.35}$$

Physically, the force $\tilde{f}$ has two components. The first one is the capillary force acting on

148 APPENDIX E. EXACT RESULTS AND LINEAR THEORY FOR ARBITRARY θ_0

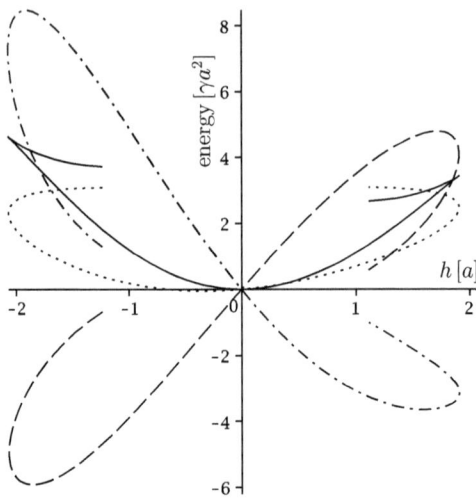

Figure E.4: Contributions to the free energy $\tilde{F}_{A0}(h)$ (solid line) defined in Eqs. (E.1)-(E.2) for the same set of parameters as in Fig.E.3: the dotted line represents the contribution numerically equal to the change in the free energy for a particle at a flat undeformable interface (Eq. (2.39)) equal to $-\pi \sin^2 \beta$; the dashed line represents the contribution $-\cos\theta_0 (S_{ol} - S_{ol,ref})$ from the solid-liquid interface; the dash-dotted line stands for the remaining contribution, i.e., $S_{lg} - S_{lg,ref} + \pi \sin^2 \beta$.

the contact line at the particle, which is given by the same expression as in the case of a flat interface (Eq. 3.16). The second component stems from the non-vanishing internal pressure of the droplet, which for an almost spherical interface is approximately equal to the Laplace pressure $2/R_0$ (and is given exactly by the Lagrange multiplier λ in Eq. (E.15)), resulting in a force of magnitude approximately $2\pi \sin^2 \beta / R_0$ directed upwards, which is of the order $O(1/R_0)$ relative to the capillary force and therefore it can be neglected for large droplets. However, for medium-sized droplets, due to the curvature of the interface, this force leads to asymmetry of $\tilde{f}(h)$ with respect to $h = 0$ (see fig. E.3(b)). Furthermore, for the same reason, unlike in the case of a flat interface, the inflection points of $\tilde{F}_{A0}(h)$, for $\theta_p = \pi/2$, does not correspond exactly to $\beta = \pi \pm \pi/4$ (see Fig. E.3(b)). From Fig. E.3(b) one can also infer that $\tilde{f}(h)$ is strongly non-linear, particularly at the metastable branches. However, the non-linearity sets in already at the stable branch for $\tilde{f} \gtrsim 2$ and for $\tilde{f} \lesssim -1$. Thus, for $f \lesssim -2$ and $f \lesssim 1$ one could expect departures from the linear perturbation theory exploited in the following Section.

Finally, we note that the contributions to the total free energy coming from the liquid-gas and the liquid-substrate interfaces are both of the same order of magnitude (see Fig. E.4) and both exhibit the aforementioned asymmetry with respect to $h = 0$.

E.2 Comparison with linear theory

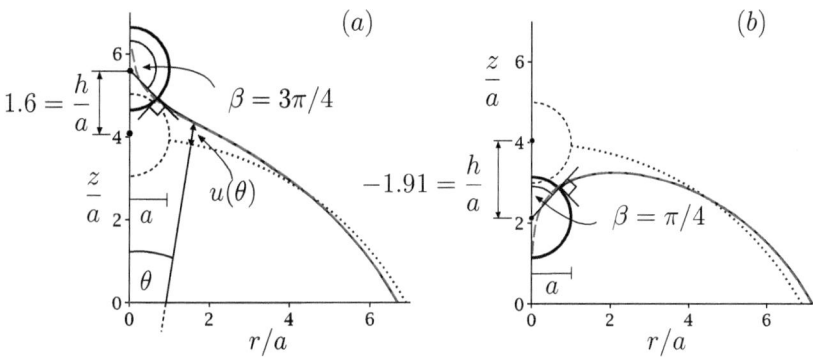

Figure E.5: Droplet shapes for $\theta_p = \pi/2$ and fixed $\theta_0 = \pi/3$ (model A), calculated by using the exact solution of the capillary equation (E.8) (black solid lines beneath the red dashed lines) for $\beta = 3\pi/4$ in (a) and for $\beta = \pi/4$ in (b), and by using the approximate solutions $R_0 \epsilon v_0(\theta)/a = u(\theta)/a$ (red dashed lines) given by Eqs. (E.36)-(E.38) for pointlike forces of amplitude $f/(\gamma a) = -\tilde{f}(h(\beta = 3\pi/4))/(\gamma a) = 2.77$ and $f/(\gamma a) = -\tilde{f}(h(\beta = \pi/4))/(\gamma a) = -3.56$, respectively (compare Fig. E.3(b)), with a cut-off at $\theta = 0.005$ and with the remaining set of parameters the same as in Fig. E.3. We note the logarithmic divergence of the approximate solutions at $\theta = 0$. The dotted lines correspond to the particle and the spherical cap shapes of the droplet in the reference configuration.

A modification of the method of images can be used for arbitrary θ_0 if the point force $f = Q\epsilon\gamma R_0$ is placed at the apex of the sessile droplet. The virtual reference droplet (for $f = 0$) is constructed as a smooth continuation of the actual droplet such that it completes the full sphere and the image of amplitude $f' = f$ is placed at the "south pole" $\theta' = \pi$. In this configuration forces acting on the union of the real and the virtual parts of the droplet are balanced. We exploit the freedom to add a constant and a term proportional to $\cos\theta$, such that the axially symmetric solution

$$v_0(\theta) = Q[G(\theta) + G(\pi - \theta) + H_0 \cos\theta + I_0] \tag{E.36}$$

conserves the volume and fulfills the given boundary conditions. Moreover, this solution fulfills Eq. (7.26) with $\pi(\Omega) = Q\,\delta(\Omega)$ and $\pi_{CM}(\Omega) = 0$ upon choosing $I_0 = -\mu/2 - 1/(4\pi)$. The value of μ, given by Eqs. (7.42) and (7.43) for $\alpha = 0$, depends on the boundary condition at the substrate.

In order to be able to make a comparison with the nonlinear theory here we take the contact line at the substrate to be free. (For the case of a pinned contact line see Appendix B in Oettel et al. (2005b).) Using the expression in Eq. (7.42) for $\alpha = 0$

yields
$$I_0 = \frac{\cos\theta_0 (1 + \cos\theta_0)}{4\pi(2 + \cos\theta_0)(1 - \cos\theta_0)}. \tag{E.37}$$

Inserting Eqs. (E.36) and (E.37) into the volume constraint $\int_0^{\theta_0} d\theta \sin\theta v_0(\theta) = 0$ one obtains
$$H_0 = \frac{1}{2\pi}\left[\ln\tan(\theta_0/2) - \frac{\cos\theta_0}{(2+\cos\theta_0)(1-\cos\theta_0)}\right]. \tag{E.38}$$

Due to $I_0(\theta_0 = \pi/2) = 0 = H_0(\theta_0 = \pi/2)$, for $\theta_0 = \pi/2$ the solution in Eq. (E.36) reduces to
$$v_0(\theta; \theta_0 = \pi/2) = Q\left[G(\theta) + G(\pi - \theta)\right] \equiv Q\, G_A(\Omega, \Omega_1 = 0), \tag{E.39}$$

which reproduces the result from Sec. 7.3.1 (Eq. (7.64)) for $\alpha = 0$. In order to express the deformation u in the dimension of length the solution in Eq. (E.36) has to be multiplied by $R_0 \epsilon = |f|/\gamma$, where $f = -\tilde{f}(h)$ must be taken from the exact solution. Particularly, by using the relation $h(\beta)$ (Eq. (E.27)), one obtains the values $f/(\gamma a) = -3.56$ and $f/(\gamma a) = 2.77$, for $\beta = \pi/4$ and $\beta = 3\pi/4$, respectively, for which the curves $z(r)$, given implicitly by $z(\theta) = [R_0 + u(\theta)]\cos\theta - z_{substr}$ and $r(\theta) = [R_0 + u(\theta)]\sin\theta$ with $u(\theta) = R_0 \epsilon v_0(\theta)$ and $z_{substr} = R_0 \cos\theta_0$, have been plotted in Fig. E.5. Apparently, even close to the edge of the stability regime the discrepancies between the perturbation theory and the exact solution are so small, that they are practically beyond the resolution of Fig. E.5 (actually $\beta = \pi/4$ corresponds already to a metastable configuration, so that it belongs to branch 6 for $h < 0$ in Fig. E.3(a)). We expect them to remain small also for the configurations without axial symmetry (which has also been justified by the numerical results in Chapter 8).

Bibliography

ABRAMOWITZ, M., & STEGUN, I. A. 1970. *Handbook of mathematical functions.* Washington, D.C.: U.S. Government Printing Office.

AIZENBERG, J., BRAUN, P. V., & WILTZIUS, P. 2000. Patterned colloidal deposition controlled by electrostatic and capillary forces. *Phys. Rev. Lett.*, **84**(13), 2997–3000.

ARCHER, A. J. 2008. Two-dimensional fluid with competing interactions exhibiting microphase separation: Theory for bulk and interfacial properties. *Phys. Rev. E*, **78**(3), 031402.

BAUSCH, A. R., BOWICK, M. J., CACCIUTO, A., DINSMORE, A. D., HSU, M. F., NELSON, D. R., NIKOLAIDES, M. G., TRAVESSET, A., & WEITZ, D. A. 2003. Grain boundary scars and spherical crystallography. *Science*, **299**(5613), 1716–1718.

BIGIONI, T. P., LIN, X. M., NGUYEN, T. T., CORWIN, E. I., WITTEN, T. A., & JAEGER, H. M. 2006. Kinetically driven self assembly of highly ordered nanoparticle monolayers. *Nature Materials*, **5**(4), 265–270.

BONN, D., OTWINOWSKI, J., SACANNA, S., GUO, H., WEGDAM, G., & SCHALL, P. 2009. Direct Observation of Colloidal Aggregation by Critical Casimir Forces. *Phys. Rev. Lett.*, **103**(15), 156101.

BOWDEN, N., TERFORT, A., CARBECK, J., & WHITESIDES, G. M. 1997. Self-assembly of mesoscale objects into ordered two-dimensional arrays. *Science*, **276**(5310), 233–235.

BRAGG, L., & NYE, J. F. 1947. A dynamical model of a crystal structure. *Proceedings of the Royal Society of London, Ser. A*, **190**(1023), 474.

BRAKKE, K. 1992. The surface evolver. *Experimental Mathematics*, **1**(2), 141–165.

BRESME, F., LEHLE, H., & OETTEL, M. 2009. Solvent-mediated interactions between nanoparticles at fluid interfaces. *J. Chem. Phys.*, **130**(21), 214711.

BRINK, D. M., & SATCHLER, G. R. 1968. *Angular Momentum.* 2 edn. Oxford: Clarendon Press.

BRINKMANN, M., KIERFELD, J., & LIPOWSKY, R. 2004. A general stability criterion for droplets on structured substrates. *J. Phys. A: Math. Gen.*, **37**(48), 11547–11573.

BROCHARD-WYART, F., DI MEGLIO, J. M., QUÉRÉ, D., & DE GENNES, P. G. 1991. Spreading of nonvolatile liquids in continuum picture. *Langmuir*, **7**, 335–338.

BROWN, A. B. D., SMITH, C. G., & RENNIE, A. R. 2000. Fabricating colloidal particles with photolithography and their interactions at an air-water interface. *Phys. Rev. E*, **62**(1), 951–960.

BUBECK, R., BECHINGER, C., NESER, S., & LEIDERER, P. 1999. Melting and reentrant freezing of two-dimensional colloidal crystals in confined geometry. *Phys. Rev. Lett.*, **82**(16), 3364–3367.

CARNAHAN, N. F., & STARLING, K. E. 1969. Equation of state for noninteracitog rigid spheres. *J. Chem. Phys.*, **51**(2), 635.

CASIMIR, H. B. G. 1948. On the attraction between two perfectly conducting plates. *Proc. K. Ned. Akad. Wet.*, **B51**, 793–795.

CHAVEZ-PAEZ, M., GONZALEZ-MOZUELOS, P., MEDINA-NOYOLA, M., & MENDEZ-ALCARAZ, J. M. 2003. Correlations among colloidal particles confined to a spherical monolayer. *J. Chem. Phys.*, **119**(14), 7461–7466.

CHEUNG, D. L., & BON, S. A. F. 2009a. Interaction of Nanoparticles with Ideal Liquid-Liquid Interfaces. *Phys. Rev. Lett.*, **102**(6), 066103.

CHEUNG, D. L., & BON, S. A. F. 2009b. Stability of Janus nanoparticles at fluid interfaces. *Soft Matter*, **5**(20), 3969–3976.

CUI, B. X., DIAMANT, H., LIN, B. H., & RICE, S. A. 2004. Anomalous hydrodynamic interaction in a quasi-two-dimensional suspension. *Phys. Rev. Lett.*, **92**(25), 258301.

DANOV, K. D., KRALCHEVSKY, P. A., NAYDENOV, B. N., & BRENN, G. 2005. Interactions between particles with an undulated contact line at a fluid interface: Capillary multipoles of arbitrary order. *J. Colloid Interface Sci.*, **287**(1), 121–134.

DE GENNES, P. G., BROCHARD-WYART, F., & QUÉRÉ, D. 2004. *Capillarity and Wetting Phenomena*. New York: Springer.

DE LAPLACE, P. S. 1805 and 1806. *Supplément au livre X du Traitée de Mechanique Céleste*. Paris: Couveier.

DEEGAN, R. D., BAKAJIN, O., DUPONT, T. F., HUBER, G., NAGEL, S. R., & WITTEN, T. A. 1997. Capillary flow as the cause of ring stains from dried liquid drops. *Nature*, **389**(6653), 827–829.

DENKOV, N. D., VELEV, O. D., KRALCHEVSKY, P. A., IVANOV, I. B., YOSHIMURA, H., & NAGAYAMA, K. 1993. 2-dimensional cristallization. *Nature*, **361**(6407), 26–26.

DIETZEL, M., & POULIKAKOS, D. 2005. Laser-induced motion in nanoparticle suspension droplets on a surface. *Phys. Fluids*, **17**(10), 102106.

DINSMORE, A. D., HSU, M. F., NIKOLAIDES, M. G., MARQUEZ, M., BAUSCH, A. R., & WEITZ, D. A. 2002. Colloidosomes: Selectively permeable capsules composed of colloidal particles. *Science*, **298**(5595), 1006–1009.

DOMÍNGUEZ, A. 2010. Capillary forces between colloidal particles at fluid interfaces. *Pages 31–59 of:* HIDALGO-ÀLVAREZ, R. (ed), *Structure and functional properties of colloidal systems*. Boca Raton: CRC Press.

DOMÍNGUEZ, A., OETTEL, M., & DIETRICH, S. 2005. Capillary-induced interactions between colloids at an interface. *J. Phys.: Condens. Matter*, **17**(45), S3387–S3392.

DOMÍNGUEZ, A., OETTEL, M., & DIETRICH, S. 2007a. Absence of logarithmic attraction between colloids trapped at the interface of droplets - Comment on "Capillary attraction of charged particles at a curved liquid interface" by Alois Wurger. *EPL*, **77**(6), 68002.

DOMÍNGUEZ, A., OETTEL, M., & DIETRICH, S. 2007b. Theory of capillary-induced interactions beyond the superposition approximation. *J. Chem. Phys.*, **127**(20), 204706.

DOMÍNGUEZ, A., OETTEL, M., & DIETRICH, S. 2008a. Force balance of particles trapped at fluid interfaces. *J. Chem. Phys.*, **128**(11), 114904.

DOMÍNGUEZ, A., FRYDEL, D., & OETTEL, M. 2008b. Multipole expansion of the electrostatic interaction between charged colloids at interfaces. *Phys. Rev. E*, **77**(2), 020401.

DOMINGUEZ, A., OETTEL, M., & DIETRICH, S. 2010. Dynamics of colloidal particles with capillary interactions. *Phys. Rev. E*, **82**(1), 011402.

EDDI, A., FORT, E., MOISY, F., & COUDER, Y. 2009. Unpredictable Tunneling of a Classical Wave-Particle Association. *Phys. Rev. Lett.*, **102**(24), 240401.

EDMONDS, A. R. 1957. *Angular Momentum in Quantum Mechanics.* 2 edn. Princeton, New Jersey: Princeton University Press.

EISENMANN, C., GASSER, U., KEIM, P., & MARET, G. 2004. Anisotropic defect-mediated melting of two-dimensional colloidal crystals. *Phys. Rev. Lett.*, **93**(10), 105702.

EVANS, R. 1979. Nature of the liquid-vapor interface and other topics in the statistical-mechanics of nonuniform, classical fluids. *Advances in Physics*, **28**(2), 143–200.

FERNANDEZ-TOLEDANO, J. C., MONCHO-JORDA, A., MARTINEZ-LOPEZ, F., & HIDALGO-ALVAREZ, R. 2004. Spontaneous formation of mesostructures in colloidal monolayers trapped at the air-water interface: A simple explanation. *Langmuir*, **20**(17), 6977–6980.

FINN, R. 1986. *Equilibrium Capillary Surfaces.* Vol. p. 1-10. New York: Springer.

FISCHER, M. 2004. Interfaces: Fluctuations, Interactions and Related Transitions. Pages 19–48 of: NELSON, D., PIRAN, T., & WEINBERG, S. (eds), *Statistical Mechanics of Membranes and Surfaces.* World Scientific.

FORET, L., & WÜRGER, A. 2004. Electric-Field Induced Capillary Interaction of Charged Particles at a Polar Interface. *Phys. Rev. Lett.*, **92**(5), 058302.

FOURNIER, J. B., & GALATOLA, P. 2002. Anisotropic capillary interactions and jamming of colloidal particles trapped at a liquid-fluid interface. *Phys. Rev. E*, **65**(3), 031601.

GAUSS, C. F. 1830. Principia generalia theoriae figurae fluidorum. *Comm. Soc. Reg. Sci. G ott. Rec.*, **7**, 65–87.

GHEZZI, F., & EARNSHAW, J. C. 1997. Formation of meso-structures in colloidal monolayers. *J. Phys.: Condens. Matter*, **9**(37), L517–L523.

GHEZZI, F., EARNSHAW, J. C., FINNIS, M., & MCCLUNEY, M. 2001. Pattern formation in colloidal monolayers at the air-water interface. *J. Colloid Interface Sci.*, **238**(2), 433–446.

GILET, T., & BUSH, J. W. M. 2009. The fluid trampoline: droplets bouncing on a soap film. *J. Fluid Mech.*, **625**(25), 167–203.

HALPERIN, B. I., & NELSON, D. R. 1978. Theory of 2-dimensional meting. *Phys. Rev. Lett.*, **41**(2), 121–124.

HAMERMESH, M. 1962. *Group theory and its application to physical problems.* London-Paris: Pergamon Press.

HELSETH, L. E., & FISCHER, T. M. 2003. Particle interactions near the contact line in liquid drops. *Phys. Rev. E*, **68**(4), 042601.

HELSETH, L. E., MURUGANATHAN, R. M., ZHANG, Y., & FISCHER, T. M. 2005. Colloidal rings in a liquid mixture. *Langmuir*, **21**(16), 7271–7275.

HERTLEIN, C., HELDEN, L., GAMBASSI, A., DIETRICH, S., & BECHINGER, C. 2008. Direct measurement of critical Casimir forces. *Nature*, **451**(7175), 172–175.

HOPKINS, P., ARCHER, A. J., & EVANS, R. 2009. Solvent mediated interactions between model colloids and interfaces: A microscopic approach. *J. Chem. Phys.*, **131**(12), 124704.

HU, D. L., & BUSH, J. W. M.. 2005. Meniscus-climbing insects. *Nature*, **437**(7059), 733–736.

HURD, A. J. 1985. The electrostatic interaction between interfacial colloidal particles. *J. Phys. A: Math. Gen.*, **18**(16), 1055–1060.

ISRAELACHVILI, J. N. 1977. Refinement of the fluid-mosaic model of membrane structure. *Biochim. Biophys. Acta*, **469**(2), 221–225.

KIM, S., & KARILLA, S. J. 1991. *Microhydrodynamics*. Boston: Butterworth-Heinemann.

KOSTERLITZ, J. M., & THOULESS, D. J. 1973. Ordering, metastablility and phase-transitions in 2 dimensional systems. *J. Phys. C: Solid State Phys.*, **6**(7), 1181–1203.

KRALCHEVSKY, P. A., & NAGAYAMA, K. 2000. Capillary interactions between particles bound to interfaces, liquid films and biomembranes. *Adv. Colloid Interface Sci.*, **85**(2-3), 145–192.

KRALCHEVSKY, P. A., & NAGAYAMA, K. 2001. *Particles at Fluid Interfaces*. Amsterdam: Elsevier.

KRALCHEVSKY, P. A., PAUNOV, V. N., DENKOV, N. D., IVANOV, I. B., & NAGAYAMA, K. 1993. Energetical and force approaches to the capillary interactions between particles attached to a liquid fluid interface. *J. Colloid Interface Sci.*, **155**(2), 420–437.

KRALCHEVSKY, P. A., PAUNOV, V. N., DENKOV, N. D., & NAGAYAMA, K. 1994. Capillary Image Forces. 1. Theory. *J. Colloid Interface Sci.*, **107**(1), 47–65.

KRALCHEVSKY, P. A., PAUNOV, V. N., & NAGAYAMA, K. 1995. Lateral capillary interaction between particles protruding from a spherical liquid layer. *J. Fluid Mech.*, **299**(25), 105–132.

KUNCICKY, D. M., & VELEV, O. D. 2008. Surface-guided templating of particle assemblies inside drying sessile droplets. *Langmuir*, **24**(4), 1371–1380.

LANGBEIN, D. 2002. *Capillary Surfaces*. 2 edn. Berlin: Springer.

LEHLE, H. 2007. *Effektive wechselwirkungen zwischen Kolloiden an Fluiden Grenzflaechen*. Universitäat Stuttgart. PhD Thesis.

LEHLE, H., OETTEL, M., & DIETRICH, S. 2006. Effective forces between colloids at interfaces induced by capillary wavelike fluctuations. *EPL*, **75**(1), 174–180.

LEHLE, H., NORUZIFAR, E., & OETTEL, M. 2008. Ellipsoidal particles at fluid interfaces. *Eur. Phys. J. E*, **26**(1-2), 151–160.

LIU, Y. H., CHEW, L. Y., & YU, M. Y. 2008. Self-assembly of complex structures in a two-dimensional system with competing interaction forces. *Phys. Rev. E*, **78**(6), 066405.

LOUDET, J. C., ALSAYED, A. M., ZHANG, J., & YODH, A. G. 2005. Capillary interactions between anisotropic colloidal particles. *Phys. Rev. Lett.*, **94**(1), 018301.

LOUDET, J. C., YODH, A. G., & POULIGNY, B. 2006. Wetting and contact lines of micrometer-sized ellipsoids. *Phys. Rev. Lett.*, **97**(1), 018304.

MADIVALA, B., FRANSAER, J., & VERMANT, J. 2009. Self-Assembly and Rheology of Ellipsoidal Particles at Interfaces. *Langmuir*, **25**(5), 2718–2728.

MEGENS, M., & AIZENBERG, J. 2003. Capillary attraction: like-charged particles at liquid interfaces. *Nature*, **424**(8), 1014.

MIKHAEL, J., ROTH, J., HELDEN, L., & BECHINGER, C. 2008. Archimedean-like tiling on decagonal quasicrystalline surfaces. *Nature*, **454**(7203), 501–504.

MORSE, D. C., & WITTEN, T. A. 1993. Droplet elasticity in weakly compressed emulsions. *Europhys. Lett.*, **22**(7), 549–555.

MURISIC, N., & KONDIC, L. 2008. Modeling evaporation of sessile drops with moving contact lines. *Phys. Rev. E*, **78**(6), 065301.

NEHARI, Z. 1975. *Conformal Mapping*. New York: Dover.

NICOLSON, M. M. 1949. The interaction between floating particles. *Proceedings of the Cambridge Philosophical Society*, **45**(2), 288–295.

NIKOLAIDES, M. G., BAUSCH, A. R., HSU, M. F., DINSMORE, A. D., BRENNER, M. P., WEITZ, D. A., & GAY, C. 2002. Electric-field-induced capillary attraction between like-charged particles at liquid interfaces. *Nature*, **420**(6913), 299–301.

OETTEL, M., & DIETRICH, S. 2008. Colloidal interactions at fluid interfaces. *Langmuir*, **24**(4), 1425–1441.

OETTEL, M., DOMÍNGUEZ, A., & DIETRICH, S. 2005a. Attractions between charged colloids at water interfaces. *J. Phys.: Condens. Matter*, **17**(32), L337–L342.

OETTEL, M., DOMÍNGUEZ, A., & DIETRICH, S. 2005b. Effective capillary interaction of spherical particles at fluid interfaces. *Phys. Rev. E*, **71**(5), 051401.

PELESKO, J. A. 2007. *Self Assembly*. Boca Raton: Chapman & Hall/CRC.

PICKERING, S. U. 1907. Emulsions. *Journal of Chemical Society*, **91**(2), 2001–2021.

PIERANSKI, P. 1980. Two-dimensional interfacial colloidal crystals. *Phys. Rev. Lett.*, **45**(7), 569–572.

PROTIERE, S., BOUDAOUD, A., & COUDER, Y. 2006. Particle-wave association on a fluid interface. *J. Fluid Mech.*, **554**(10), 85–108.

PRUDNIKOV, A. P., BRYCHKOV, Y. A., & MARICHEV, O. I. 1986a. *Integrals and Series*. 2 edn. Vol. 2. New York: Gordon and Breach.

PRUDNIKOV, A. P., BRYCHKOV, Y. A., & MARICHEV, O. I. 1986b. *Integrals and Series*. 2 edn. Vol. 1. New York: Gordon and Breach.

Rosso, R., & Virga, E. G. 2003. General stability criterion for wetting. *Phys. Rev. E*, **68**(1), 012601.

Roth, R., Gotzelmann, B., & Dietrich, S. 1999. Depletion forces near curved surfaces. *Phys. Rev. Lett.*, **83**(2), 448–451.

Rowlinson, J. S., & Widom, B. 2002. *Molecular Theory of Capillarity*. New York: Dover.

Ruiz-Garcia, J., Gamez-Corrales, R., & Ivlev, B. I. 1998. Formation of two-dimensional colloidal voids, soap froths, and clusters. *Phys. Rev. E*, **58**(1), 660–663.

Sangani, A. S., Lu, C. H., Su, K. H., & Schwarz, J. A. 2009. Capillary force on particles near a drop edge resting on a substrate and a criterion for contact line pinning. *Phys. Rev. E*, **80**(1), 011603.

Schimmele, L., Napiorkowski, M., & Dietrich, S. 2007. Conceptual aspects of line tensions. *J. Chem. Phys.*, **127**(16), 164715.

Sear, R. P., Chung, S. W., Markovich, G., Gelbart, W. M., & Heath, J. R. 1999. Spontaneous patterning of quantum dots at the air-water interface. *Phys. Rev. E*, **59**(6), R6255–R6258.

Stamou, D., Duschl, C., & Johannsmann, D. 2000. Long-range attraction between colloidal spheres at the air-water interface: The consequence of an irregular meniscus. *Phys. Rev. E*, **62**(4), 5263–5272.

Tarazona, P., Cuesta, J. A., & Martínez-Ratón, Y. 2008. Density functional theories of hard particle systems. Pages 247–341 of: Mulero, A. (ed), *Theory and Simulation of Hard-Sphere Fluids and Related Systems*. Berlin/Heidelberg: Springer.

Tasinkevych, M,, & Dietrich, S. 2006. Complete wetting of nanosculptured substrates. *Phys. Rev. Lett.*, **97**(10), 106102.

Tasinkevych, M,, & Dietrich, S. 2007. Complete wetting of pits and grooves. *Eur. Phys. J. E*, **23**(1), 117–128.

van Blaaderen, A,, Ruel, R,, & Wiltzius, P. 1997. Template-directed colloidal crystallization. *Nature*, **385**(6614), 321–324.

van Nierop, E. A., Stijnman, M. A., & Hilgenfeldt, S. 2005. Shape-induced capillary interactions of colloidal particles. *EPL*, **72**(4), 671–677.

Vella, D., & Mahadevan, L. 2005. The "Cheerios effect". *Americal Journal of Physics*, **73**(9), 817–825.

Viveros-Mendez, P. X., Mendez-Alcaraz, J. M., & Gonzalez-Mozuelosa, P. 2008. Two-body correlations among particles confined to a spherical surface: Packing effects. *J. Chem. Phys.*, **128**(1), 014701.

WEEKS, J. D., CHANDLER, D., & ANDERSEN, H. C. 1971. Role of repulsive forces in determining equilibrium structure of simple liquids. *J. Chem. Phys.*, **54**(12), 5237.

WHITESIDES, G. M., & GRZYBOWSKI, B. 2002. Self-assembly at all scales. *Science*, **295**(5564), 2418–2421.

WIGNER, E. P. 1959. *Group theory and its application to the quantum mechanics of atomic spectra.* 2 edn. London: Academic Press.

WÜRGER, A. 2006a. Capillary attraction of charged particles at a curved liquid interface. *EPL*, **75**(6), 978–984.

WÜRGER, A. 2006b. Curvature-induced capillary interaction of spherical particles at a liquid interface. *Phys. Rev. E*, **74**(4), 041402.

YOUNG, A. P. 1978. Theory of phase-transition in 2-dimensional planar spin model. *J. Phys. C: Solid State Phys.*, **11**(11), L453–L455.

YOUNG, T. 1805. An essay on the cohesion of fluids. *Philosophical Transactions of the Royal Society of London*, **95**, 65–87.

ZAHN, K., & MARET, G. 2000. Dynamic criteria for melting in two dimensions. *Phys. Rev. Lett.*, **85**(17), 3656–3659.

ZAHN, K., LENKE, R., & MARET, G. 1999. Two-stage melting of paramagnetic colloidal crystals in two dimensions. *Phys. Rev. Lett.*, **82**(13), 2721–2724.

I want morebooks!

Buy your books fast and straightforward online - at one of world's fastest growing online book stores! Environmentally sound due to Print-on-Demand technologies.

Buy your books online at
www.morebooks.shop

Kaufen Sie Ihre Bücher schnell und unkompliziert online – auf einer der am schnellsten wachsenden Buchhandelsplattformen weltweit! Dank Print-On-Demand umwelt- und ressourcenschonend produziert.

Bücher schneller online kaufen
www.morebooks.shop

KS OmniScriptum Publishing
Brivibas gatve 197
LV-1039 Riga, Latvia
Telefax: +371 686 204 55

info@omniscriptum.com
www.omniscriptum.com

Printed by Books on Demand GmbH, Norderstedt / Germany